HUMAN FACTORS METHODS AND ACCIDENT ANALYSIS

Human Factors Methods and Accident Analysis
Practical Guidance and Case Study Applications

PAUL M. SALMON
Monash University, Australia

NEVILLE A. STANTON
University of Southampton, UK

MICHAEL LENNÉ
Monash University, Australia

DANIEL P. JENKINS
Sociotechnic Solutions, UK

LAURA RAFFERTY
University of Southampton, UK

GUY H. WALKER
Heriot-Watt University, UK

ASHGATE

© Paul M. Salmon, Neville A. Stanton, Michael Lenné, Daniel P. Jenkins, Laura Rafferty and Guy H. Walker 2011

All rights reserved. No part of this publication may be reproduced, stored in a retrieval system or transmitted in any form or by any means, electronic, mechanical, photocopying, recording or otherwise without the prior permission of the publisher.

Paul M. Salmon, Neville A. Stanton, Michael Lenné, Daniel P. Jenkins, Laura Rafferty and Guy H. Walker have asserted their moral right under the Copyright, Designs and Patents Act, 1988, to be identified as the authors of this work.

Published by
Ashgate Publishing Limited
Wey Court East
Union Road
Farnham
Surrey, GU9 7PT
England

Ashgate Publishing Company
Suite 420
101 Cherry Street
Burlington
VT 05401-4405
USA

www.ashgate.com

British Library Cataloguing in Publication Data
Human factors methods and accident analysis : practical guidance and case study applications.
  1. Accident investigation. 2. Accident investigation--Case studies. 3. Human behavior models.
  I. Salmon, Paul M.
  363.1'065-dc22

Library of Congress Cataloging-in-Publication Data
Human factors methods and accident analysis : practical guidance and case study applications / by Paul M. Salmon ... [et al.].
    p. cm.
  Includes bibliographical references and index.
  ISBN 978-1-4094-0519-1 (hbk) -- ISBN 978-1-4094-0520-7 (ebook)
  1. Accidents. 2. Accidents--Research. 3. Accident investigation.
  4. Errors--Evaluation. I. Salmon, Paul M.
  HV675.H78 2011
  363.1'065--dc23
                                                                                  2011024619

ISBN: 978-1-4094-0519-1 (hbk)
ISBN: 978-1-4094-0520-7 (ebk)

Printed and bound in Great Britain by the
MPG Books Group, UK

Contents

List of Figures	*vii*
List of Tables	*ix*
About the Authors	*xi*
Acknowledgements	*xv*
Acronyms	*xvii*
Preface	*xix*

1 Accidents, Accident Causation Models and Accident Analysis Methods   1

2 Human Factors Methods for Accident Analysis   7

3 AcciMap: Lyme Bay Sea Canoeing and Stockwell Mistaken Shooting Case Studies   83

4 The Human Factors Analysis and Classification System: Australian General Aviation and Mining Case Studies   97

5 The Critical Decision Method: Retail Store Worker Injury Incident Case Study   123

6 Propositional Networks: Challenger II Tank Friendly Fire Case Study   133

7 Critical Path Analysis: Ladbroke Grove Case Study   143

8 Human Factors Methods Integration: Operation Provide Comfort Friendly Fire Case Study   151

9 Discussion   177

References   181
Index   191

List of Figures

1-1	Reproduction of Reason's Swiss cheese accident causation model	3
1-2	Rasmussen's risk management framework along with migration of work practices models	3
2-1	Generic accident analysis procedure	9
2-2	Victorian fatal road traffic crash data mapped onto accident causation frameworks	11
2-3	Hillsborough disaster AcciMap	27
2-4	*Herald of Free Enterprise* incident fault tree extract	32
2-5	HFACS taxonomies overlaid on Reason's Swiss cheese model	33
2-6	HFACS analysis of Lyme Bay sea canoeing tragedy	39
2-7	Generic control structure model	42
2-8	Extract of systems dynamics model of Walkerton water contamination incident	43
2-9	Mangatepopo incident: basic control structure diagram	46
2-10	Example failures in control structures between field manager and instructor, and between instructor and students	47
2-11	Example social network diagram for Tenerife air disaster communications	50
2-12	Potential rooftop rescue vignette	53
2-13	NYPD North Tower evacuation vignette; figure shows the social network involved and the situation awareness exchanges between parties	54
2-14	FDNY North Tower evacuation vignette; figure shows the social network involved and the situation awareness exchanges between parties	55
2-15	Example relationships between concepts	58
2-16	*Herald of Free Enterprise* incident; actual and prevention networks	62
2-17	Task sequence/dependency chart for ATM example	66
2-18	Network of networks approach to analysing collaborative activities; figure shows example representations of each network, including hierarchical task analysis (task network), social network analysis (social network) and propositional network (knowledge network) representations	77
3-1	Lyme Bay tragedy AcciMap	85
3-2	AcciMap (coded by phase, links indicate causal relationships, weak causal links shown as dotted lines)	93
4-1	Signal detection paradigm matrix	106
4-2	Number of significant incidents by event type	112

4-3	Number of significant incidents by mining activity	112
4-4	Associations between HFACS codes at each level	115
6-1	Hierarchical organisation of the system	134
6-2	Overall incident propositional network	137
6-3	Propositional network showing DSA-related failures	138
7-1	Ladbroke Grove incident track layout schematic	144
7-2	Signaller's work station	145
7-3	CPA representation	148
7-4	CPA analysis of signalman's activities post SPAD	149
8-1	Network of networks approach	152
8-2	Network of network approach overlaid with methods applied during case study	153
8-3	Command structure of the OPC	155
8-4	Task network for the actual and ideal scenarios	160
8-5	Actual scenario social network	161
8-6	Ideal scenario social network	161
8-7	Sociometric status for actual and ideal scenarios	163
8-8	Centrality values for actual and ideal scenarios	164
8-9	Usage of different communications media	166
8-10	Coordination demands analysis coordination dimensions	168
8-12	Propositional network representing systems situation awareness at 8.22 in the ideal scenario	169
8-11	Propositional network representing systems situation awareness at 8.22 in the actual scenario	170
9-1	Systems analysis capability of accident analysis methods; graph shows extent to which each method considers contributory factors at each of the levels described in Rasmussen's framework.	179

List of Tables

1-1	Events and outcomes	1
2-1	Human factors accident analysis methods summary table	13
2-2	Accident analysis critical decision method probes	18
2-3	Retail store worker injury accident CDM transcript	21
2-4	Unsafe acts level external error mode taxonomies	34
2-5	STAMP's control failure categories	43
2-6	Social network agent association matrix example	49
2-7	Communication failure types	56
2-8	Input/output modalities for ATM example	66
2-9	Critical path calculation table	67
2-10	Extracts from TRACEr's task error and performance shaping factors taxonomy	71
2-11	TRACEr external error mode taxonomy	72
2-12	Internal error modes and psychological error mechanisms	73
4-1	Aircraft types involved in incidents analysed	98
4-2	Summary of flight purpose	98
4-3	Phase of operation	100
4-4	Outcome of incident	100
4-5	Effect on flight	101
4-6	Frequency and percentage of HFACS codes identified in GA incidents	103
4-7	HFACS analysis by purpose of flight	105
4-8	Agreement in error classifications between assessors	106
4-9	Comparison of case study findings with selected HFACS aviation analyses	108
4-10	Frequency and percentage of HFACS codes identified in mining incidents	113
4-11	Odds ratios for the associations between codes from each HFACS level	114
4-12	Summary of recommendations for countermeasures/incident prevention strategies proposed on basis of HFACS analysis	121
5-1	CDM probes	125
5-2	Example CDM interview transcript	126
5-3	Summary of CDM interview findings	127
6-1	Information element failure types and causal factors	139
7-1	Estimates of activity times from the literature on HCI	147
8-1	Table of acronyms	154
8-2	Timeline of events	156
8-3	Actual Scenario CUD analysis extract	166
8-4	CDA teamwork taxonomy	167

8-5	Key information elements for both the actual and ideal scenarios	171
8-6	Actual scenario key information elements per scenario phase	172
8-7	Ideal scenario key information elements per scenario phase	173
8-8	Operation loading table for actual scenario	174
8-9	Operation loading table for ideal scenario	175

About the Authors

DR PAUL M. SALMON

Human Factors Group, Monash Injury Research Institute,
Accident Research Centre and Disaster Resilience Unit,
Clayton Campus, Monash University, Victoria 3800, Australia
paul.salmon@monash.edu

Paul Salmon is a Senior Research Fellow within the Human Factors Group at the Monash Injury Research Institute and holds an Australian National Health and Medical Research Council (NHMRC) post-doctoral training fellowship in the area of Public Health. Paul has almost a decade of experience in applied Human Factors research in a number of domains, including the military, aviation, and road and rail transport and has co-authored eight books, over 60 peer-reviewed journal articles, and numerous conference articles and book chapters. Paul was recently awarded the 2007 Royal Aeronautical Society Hodgson Prize for a co-authored paper in the society's *Aeronautical Journal*, and, along with his colleagues from the Human Factors Integration Defence Technology Centre (HFI DTC), the 2008 Ergonomics Society's President's Medal. Paul was also recently named as a finalist in the Scopus Australian young researcher of the year award for Humanities and Social Sciences.

PROFESSOR NEVILLE A. STANTON

Transportation Research Group, University of Southampton
School of Civil Engineering and the Environment
Highfield, Southampton, SO17 1BJ
n.stanton@soton.ac.uk

Professor Stanton holds a Chair in Human Factors in the School of Civil Engineering and the Environment at the University of Southampton. He has published over 150 peer-reviewed journal papers and 20 books on Human Factors and Ergonomics. In 1998, he was awarded the Institution of Electrical Engineers Divisional Premium Award for a co-authored paper on Engineering Psychology and System Safety. The Ergonomics Society awarded him the Otto Edholm medal in 2001 and The President's Medal in 2008 for his contribution to basic and applied ergonomics research. In 2007, The Royal Aeronautical Society awarded him the Hodgson Medal and Bronze Award with colleagues for their work on flight deck safety. Professor Stanton is an editor of the journal *Ergonomics* and is on the editorial boards of *Theoretical Issues in Ergonomics Science* and the *International Journal of Human Computer Interaction*. Professor Stanton is a Fellow and Chartered Occupational Psychologist registered with The British Psychological Society, and a Fellow of The Ergonomics Society. He has a BSc (Hons) in Occupational Psychology from the University of Hull, an MPhil in Applied Psychology and a PhD in Human Factors from Aston University in Birmingham.

DR MICHAEL G. LENNÉ

Human Factors Group Leader,
Monash University Accident Research Centre, Building 70,
Clayton Campus, Monash University, Victoria 3800, Australia
Michael.lenne@.monash.edu

Michael Lenné leads the Human Factors Team at the Monash University Accident Research Centre, and was recently appointed as Adjunct Associate Professor and Deputy Director of the Curtin-Monash Accident Research Centre in Perth, Western Australia. Michael has a PhD in Experimental Psychology and his research over the past 18 years has been centred on the measurement of human performance using simulation in road, rail, and military aviation and maritime settings. While simulation remains a strong interest personally, and also to the team of scientists he leads, the centre's recent acquisition of on-road test vehicles affords new means of measuring driver performance. His current research uses simulation and on-road testing to study the impacts of infrastructure design on human performance and error, the impacts of in-vehicle systems on driver performance and distraction, and the influences of alcohol and other drugs on driving performance. He also has expertise in human error, accident investigation and system safety, brought about through the development of accident investigation and reporting programs with the aviation industry, and the analysis of data in both transportation and non-transportation domains using human factors methods.

DR DANIEL P. JENKINS

Sociotechnic Solutions Ltd, 2 Mitchell Close, St Albans, Herts, UK.
dan@sociotechnic.com

Dan Jenkins is a freelance human factors engineer and Director of Sociotechnic Solutions Limited. Dan started his career as an automotive engineer, graduating in 2004, with an M.Eng (Hons) in Mechanical Engineering and Design, receiving the University Prize for the highest academic achievement in the school. During his time in the car industry, Dan developed a keen interest in ergonomics and human factors. In 2005, Dan returned to Brunel University taking up the full-time role of Research Fellow in the Ergonomics Research Group. Dan studied part time for his PhD in Human Factors and Interaction Design, graduating in 2008 and receiving the Hamilton Prize for the Best Viva in the School of Engineering and Design. In 2009, Dan started his own consultancy, which has allowed him to gain industrial experience across a wide range of domains. Dan has developed experience of applied research in various domains including defence, nuclear facilities, automotive, submarines, aviation, policing, and control room design. Dan has co-authored eight books and over 40 peer-reviewed journal papers, alongside numerous conference articles and book chapters. Dan and his colleagues were awarded the Ergonomics Society's President's Medal in 2008 for their contribution to basic and applied ergonomics research.

DR LAURA RAFFERTY

Transportation Research Group, University of Southampton
School of Civil Engineering and the Environment
Highfield, Southampton, SO17 1BJ
l.rafferty@soton.ac.uk

Laura Rafferty is a research fellow in the Transportation Research Group at the University of Southampton and holds a PhD in Human Factors and a BSc in Psychology (Hons). Laura has over five years experience in applied Human Factors research in the military covering a range of applications, including accident causation and analysis, fratricide, naturalistic decision making in teams, and Human Factors methods. Laura has authored and co-authored various publications, including three books and numerous peer-reviewed journal articles and conference papers.

DR GUY H. WALKER

School of the Built Environment, Heriot-Watt University,
Edinburgh, UK, EH14 4AS
G.H.Walker@hw.ac.uk

Guy Walker is a lecturer in the School of the Built Environment at Heriot-Watt University, Edinburgh, and his research focuses on human factors issues in infrastructure and transport. He is a recipient, with his colleagues, of the Ergonomics Society's President's Medal for original research. He is also author/co-author of nine books on diverse topics in human factors, including a major text on human factors methods, and is author/co-author of over 50 international peer-reviewed journal papers.

Acknowledgements

We are indebted to many colleagues who have been involved in either the studies described in this book or the preparation and review of the manuscript itself. Special thanks go to all of those colleagues who were involved in the research underpinning the various case studies presented, including Chris Baber, Missy Rudin-Brown, Eve Mitsopoulos-Rubens, Amy Williamson, Charles Liu, Margaret Trotter, Elizabeth Varvaris, Mike Regan, Narelle Haworth, Nicola Fotheringham, Karen Ashby, and Michael Fitzharris. We are also grateful to the various organisations that had the foresight and initiative to initiate and fund the various research programmes described and also to the many subject matter experts who acted over and above the call of duty to assist in data collection and analysis activities, and who provided valuable insight when conducting analyses and reviewing analysis outputs. Finally, we wish to thank Kerri Salmon for reviewing the final draft of this manuscript, and also Miranda Cornelissen for her helpful comments throughout.

Paul Salmon's contribution to this book was funded through his Australian National Health and Medical Research Council training fellowship and also a small grant from the Monash University Accident Research Centre strategic development programme fund.

Acronyms

The following is a reference list of acronyms used within this book.

Acronym	Explanation
ACE	Air Command Element
ACO	Air Control Order
ASFA	Aviation Safety Foundation Australia
ASO	Air Surveillance Officer
ATO	Air Tasking Order
AWACS	Airborne Warning and Control System
BBC	British Broadcasting Corporation
BG	Battle Group
BH	Black Hawk
CCTV	Closed-Circuit Television
CDA	Coordination Demands Analysis
CDM	Critical Decision Method
CFAC	Combined Forces Air Component
CI	Confidence Intervals
CPA	Critical Path Analysis
CREAM	Cognitive Reliability and Error Analysis Method
CTA	Cognitive Task Analysis
CTF	Combined Task Force
CUD	Communications Usage Diagram
DSA	Distributed Situation Awareness
DSO	Designated Senior Officer
EAST	Event Analysis of Systemic Teamwork
EEM	External Error Mode
EMS	Emergency Medical Services
FDNY	Fire Department New York
FRCP	Fatal Risk Control Protocol
FTA	Fault Tree Analysis
GA	General Aviation
GPI	Gun Position Indicator

HESH	High Explosive Squash Head
HFACS	Human Factors Analysis and Classification System
HQ	Headquarters
HSE	Health and Safety Executive
HTA	Hierarchical Task Analysis
ICAM	Incident Cause Analysis Method
IEM	Internal Error Mode
IFF	Identification Friend or Foe
IPCC	Independent Police Complaints Commission
JCdM	Jean Charles de Menezes
JOIC	Joint Operations Intelligence Centre
MCC	Military Coordination Centre
MoD	Ministry of Defence
NYPD	New York Police Department
OPC	Operation Provide Comfort
OR	Odds Ratio
OSD	Operation Sequence Diagram
PEM	Psychological Error Mechanism
PSF	Performance Shaping Factor
ROE	Rules of Engagement
SD	Senior Director
SHERPA	Systemic Human Error Reduction and Prediction Approach
SME	Subject Matter Expert
SNA	Social Network Analysis
SPINS	Special Instructions
STAMP	Systems Theoretic Accident Modelling and Processes
TAOR	Tactical Area of Responsibility
TRACEr	Technique for the Retrospective Analysis of Cognitive Errors
VID	Visual Identification
WESTT	Workload, Error, Situation awareness, Time and Teamwork

Preface

Accidents, accident causation, and accident prevention remain key themes within Human Factors and Ergonomics research efforts worldwide. At the time of writing this book, well over 2,000 peer-reviewed journal articles containing the term 'accident' in either their title, abstract or keywords, had been published since the 1970s in what can be considered to be the core Human Factors and Ergonomics journals (*Ergonomics, Human Factors, Applied Ergonomics, Theoretical Issues in Ergonomics Science, Safety Science,* and *Accident Analysis and Prevention*). Further, a scan of the academic literature reveals recent accident-focused applications across a wide range of domains, including road transport (Clarke et al. 2010), aviation (Griffin et al. 2010), healthcare (Harms-Ringdahl 2009), shipping (Celik and Cebi 2009), rail (Baysari et al. 2009), outdoor education (Salmon, Williamson et al. 2010), mining (Patterson and Shappell 2010), construction (Wu et al. 2010), armed police response (Jenkins et al. 2010), the military (Rafferty et al. 2010), the food industry (Jacinto et al. 2009) farming (Cassano-Piche et al. 2009) and dynamite production (Le Coze 2010).

Despite this proliferation of accident-focused research, it is generally acknowledged that our understanding of accidents remains incomplete and that accidents will continue to occur within complex sociotechnical systems (Hollnagel 2004). This arises not as a function of poor research; rather it reflects the evolving, probabilistic complexity inherent in how accidents unfold. The methods that Human Factors researchers and practitioners, accident investigators, and safety practitioners use to analyse or investigate accidents are therefore critical to aid our understanding of the underlying causes as well as indicating where system safety may be improved. Since it is acknowledged that accidents will continue to occur, it is highly important that as much as is possible is learnt from them when they do (Reason 2008). James Reason, one of the great minds in this area, talks of the need for organisations to learn as much as is possible from accident and near-miss data, likening successful organisations in this respect to nervous, twitchy squirrels who constantly monitor their surroundings for potential enemies (Reason 2008). In short, without appropriate accident analysis methods, our understanding of accidents, and our ability to prevent them, will remain limited.

Thankfully, numerous methods exist for accident analysis purposes, with a large number, unsurprisingly, emerging from the discipline of Human Factors. Human Factors is concerned with the study of human performance in sociotechnical systems or 'the scientific study of the relationship between man and his working environment' (Murrell 1965). Broadly, the discipline of Human Factors exists to understand and enhance the safety and efficiency of sociotechnical systems. Accordingly, an array of methods for describing, understanding, and evaluating the performance of complex sociotechnical systems are used by Human Factors researchers and practitioners. These structured methods form a major part of our discipline, with a recent review identifying well over a hundred reported in academic texts (Stanton et al. 2005). They cover a variety of issues, ranging from individual operator concepts, such as situation awareness, decision-making, and workload, to team concepts, such as teamwork, coordination and collaboration, to systems concepts, such as accidents, safety culture, training and system design. Although many different methods exist, developed and applied for very different purposes, a large portion of them can be applied to the analysis of accidents. Some of these are obvious, since they were developed specifically for this purpose; however, some are not so obvious, since they were developed for other purposes, and yet each has their own utility in offering unique perspectives on accidents and their causation. The purpose of this book is to present detailed guidance on a sub-set of Human Factors methods that the authors have previously used for accident analysis purposes in a range of domains.

1.1 READERSHIP AND STRUCTURE

This book is written for both practitioners and researchers, including students, who are involved or wish to be involved in accident analysis and investigation. The intention is that the reader can use the book to select an appropriate accident analysis methodology to suit their analysis needs, and then use the practical guidance and case study examples provided to see how the method works and then apply the method effectively.

This book has therefore been constructed so that readers can read the chapters non-linearly and independently from one another. For those not familiar with accident causation models, Chapter 1 presents a brief overview of accident causation models and accident analysis methods in general. This is not intended to be groundbreaking or offer any new perspective on accident causation; rather it is offered as a general overview of the current status quo within the Human Factors literature. Chapter 2 provides a short introduction to accident analysis methods, and then presents detailed information and guidance on a sub-set of Human Factors methods that have been, or can be, applied for accident analysis purposes. Each method presented is described using the following Human Factors methods description criteria adapted from criteria that we have found useful in the past (e.g. Salmon, Stanton et al. 2010; Stanton et al. 2004; Stanton et al. 2005):

1. Name and acronym – the name of the method and its associated acronym;
2. Background and applications – provides a short introduction to the method, including a brief overview of the method and its origins and development;
3. Domain of application – describes the domain that the method was originally developed for and applied in and any alternative domains in which the method has since been applied;
4. Applications for accident analysis/investigation purposes – denotes whether the method has previously been applied for accident analysis and investigation purposes;
5. Procedure and advice – describes a step-by-step procedure for applying the method as well as general points of advice;
6. Flowchart – presents a flowchart depicting the procedure that analysts should follow when applying the method;
7. Advantages – lists the main advantages associated with using the method for accident analysis purposes;
8. Disadvantages – lists the main disadvantages associated with using the method for accident analysis purposes;

9. Example output – presents an example, or examples, of the outputs derived from applications of the method in question;
10. Related methods – lists any closely related methods, including contributory methods that should be applied in conjunction with the method, other methods to which the method acts as an input, and similar methods.
11. Approximate training and application times – estimates of the training and application times are provided to give the reader an idea of the commitment required when using the method;
12. Reliability and validity – any evidence, published in the academic literature, on the reliability or validity of the method is cited;
13. Tools needed – describes any additional tools (e.g. software packages, video and audio recording devices, flipcharts) required when using the method; and
14. Recommended texts – a bibliography lists recommended further reading on the method and the surrounding topic area.

The methods criteria approach works on three levels. First, it provides a detailed overview of the method in question. Second, it provides researchers and practitioners with some of the information that they may require when selecting an appropriate method to use for a particular analysis effort (e.g. associated methods, example outputs, flowcharts, training and application times, tools needed, and recommended further reading) and, third, it provides detailed guidance, in a step-by-step format, on how to apply the chosen method.

Following Chapter 2, each chapter presents case study accident analysis applications involving an application by the authors of one of the methods described in Chapter 2. Each case study is presented in a structured format that is intended to give the reader a detailed overview of why and how the chosen method was applied, and also the outputs and conclusions derived from the analysis. Following seven case studies, Chapter 8 presents a methods integration case study, whereby a framework of integrated Human Factors methods was used for accident analysis purposes. An overview of each case study chapter is given below.

Chapter 3: AcciMap: led outdoor activity and armed police response case studies. Chapter 3 presents two case studies involving the popular AcciMap accident analysis method (Rasmussen 1997). The first case study involved the use of the AcciMap method to describe the system-wide failures involved in the Lyme Bay (off the south coast of England) sea canoeing tragedy. The second case study involved the application of the AcciMap method to the analysis of the Stockwell (south London, UK) Jean Charles De Menezes shooting.

Chapter 4: The Human Factors Analysis and Classification System: Australian general aviation and mining case studies. Chapter 4 presents two case studies involving the Human Factors Analysis and Classification System (HFACS; Wiegmann and Shappell 2003). The first case study involved an application of HFACS to the analysis of Australian general aviation incidents derived from insurance databases. The second case study involved an application of HFACS to the analysis of 263 significant coal mining incidents occurring at a major Australian mining company during 2007–2008.

Chapter 5: The Critical Decision Method: retail store worker injury incident case study. Chapter 5 presents a case study involving the application of the Critical Decision Method cognitive task analysis approach (CDM; Klein et al. 1989) for investigating decision-making prior to accident involvement in the retail domain. The CDM was used to investigate the factors influencing retail store workers' decision-making prior to involvement in 49 injury-causing accidents.

Chapter 6: Propositional Networks: friendly fire case study. Chapter 6 presents a case study involving the application of the propositional networks analysis method (Salmon et al. 2009) to the analysis of a military friendly fire incident. The propositional networks method was used to describe the situation awareness-related failures involved in the Gulf war conflict Basra Challenger II tank friendly fire incident.

Chapter 7: Critical Path Analysis: Ladbroke Grove case study. Chapter 7 presents a case study involving the application of the Critical Path Analysis (CPA) modelling approach to the analysis of the UK Ladbroke Grove rail incident. CPA was used to model temporally the signaller's response to track occupation warnings prior to the incident.

Chapter 8: Methods Integration Case study: in Chapter 8, a case study involving the application of an integrated framework of different Human Factors methods to analyse the Operation Provide Comfort Black Hawk friendly fire incident is presented. The purpose of this case study is to demonstrate how Human Factors methods can usefully be applied together in an integrated manner in order to analyse accidents.

Chapter 9: Discussion: in Chapter 9 a short discussion is presented on the utility of the accident analysis methods discussed. The capability of each method for identifying causal factors across the overall complex sociotechnical system is discussed through mapping the outputs of each method onto a popular accident causation framework.

1
Accidents, Accident Causation Models and Accident Analysis Methods

1.1 INTRODUCTION

A topic of interest for many years (e.g. Heinrich 1931), various perspectives on accident causation are presented in the academic literature. Whilst not wishing to enter into debate over which are the most appropriate accident causation models to emerge from the discipline of Human Factors, it is worthwhile to orient the reader by providing a summary of some of the more prominent perspectives available. First and foremost, however, clarification is required over what we mean by the term 'accident'. Hollnagel (2004) presents a neat account of the etymology surrounding the term accident, and goes on to define an accident as follows: 'a short, sudden, and unexpected event or occurrence that results in an unwanted and undesirable outcome' (Hollnagel 2004, 5).

Hollnagel (2004) further points out the event must be short rather than slowly developing and must be sudden in that it occurs without prior warning. Usefully, to minimise confusion further, Hollnagel also goes on to distinguish between accidents, bad luck, misfortune, good luck, and achievement or goal fulfilment. This distinction is presented in Table 1-1, which shows that although there are two situations in which the outcome is unwanted (i.e. an accident or bad luck/misfortune), only those in which the event is unexpected or unpredictable should be labelled accidents.

Table 1-1   Events and outcomes

	Outcome is unwanted	Outcome is desired
Event is unexpected or unpredictable	Accident	Stroke of good luck
Event is expected or predictable	Bad luck, misfortune	Achievement, goal fulfilment

Source: Hollnagel 2004

1.2 MODELS OF ACCIDENT CAUSATION

Various models of accident causation exist (e.g. Heinrich 1931; Leveson 2004; Perrow 1999; Rasmussen 1997; Reason 1990) each of which engender their own approach to accident analysis. Generally speaking, the view on accident causation has evolved somewhat over the past century, with an early focus on hardware or equipment failures being superseded by increased scrutiny on the unsafe acts or 'human errors' made by operators, following which failures in the wider organisational system became the prominent focus during the late 1980s and early 1990s. It is now widely accepted that the accidents which occur in complex sociotechnical systems are caused by a range of interacting human and systemic factors (e.g. Reason 1990).

Although some disagree with the classification (e.g. Reason 2008), Hollnagel (2004) distinguishes between three types of accident causation model: sequential, epidemiological, and systemic accident models. Sequential models, characterised by Heinrich's (1931) domino theory accident causation model, view accidents simply as the result of a sequence of linear events, with the last event being the accident itself. Epidemiological models, characterised by Reason's 'Swiss cheese' model (although Reason himself disagrees with this, Reason 2008), view accidents much like the spreading of disease (Hollnagel 2004), and describe the combination of latent conditions present in the system for some time and their role in unsafe acts made by operators at the so-called 'sharp end' (Reason 1997; Hollnagel 2004). Finally, systemic models, as characterised by Leveson's Systems Theoretic Accident Modelling and Processes (STAMP; Leveson 2004) model, focus on the performance of the system as a whole, as opposed to linear cause effect relationships or epidemiological factors within the system (Hollnagel 2004). Under this approach, accidents are treated as an emergent property of the overall sociotechnical system.

In the following section we will briefly outline some of the more prominent accident causation models. Regardless of classification, undoubtedly the most popular and widely applied model is Reason's (1990) Swiss cheese model (Figure 1-1). Its popularity is such that it has driven development of various accident analysis methods, e.g. the HFACS aviation accident analysis method (Wiegmann and Shappell 2003) and has also been commonly applied itself as an accident analysis framework (e.g. Lawton and Ward 2005). Reason's model describes the interaction between system-wide 'latent conditions' (e.g. poor designs and inadequate equipment, inadequate supervision, manufacturing defects, maintenance failures, inadequate training, poor procedures) and unsafe acts made by human operators and their contribution to accidents. Importantly, rather than focus on the front line or so called 'sharp end' of system operation, Reason's model describes how these latent conditions reside across all levels of the organisational system (e.g. higher supervisory and managerial levels). According to the model, each organisational level has defences, such as protective equipment, rules and regulations, training, checklists and engineered safety features, which are designed to prevent the occurrence of occupational accidents. Weaknesses in these defences, created by latent conditions and unsafe acts, create 'windows of opportunity' for accident trajectories to breach the defences and cause an accident.

Although not enjoying the high levels of exposure and popularity of Reason's model, other accident causation models are also prominent. Rasmussen's risk management framework (Rasmussen 1997, see Figure 1-2), e.g. is also popular, and has also driven the development of accident analysis methods (e.g. AcciMap; Rasmussen 1997). Rasmussen's framework is based on the notion that accidents are shaped by the activities of people

CHAPTER 1 • ACCIDENTS, ACCIDENT CAUSATION MODELS, AND
ACCIDENT ANALYSIS METHODS

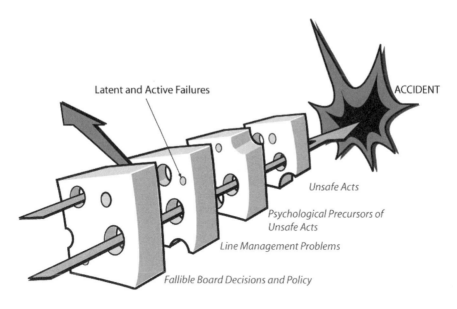

Figure 1-1   Reproduction of Reason's Swiss cheese accident causation model (adapted from Reason 1990)

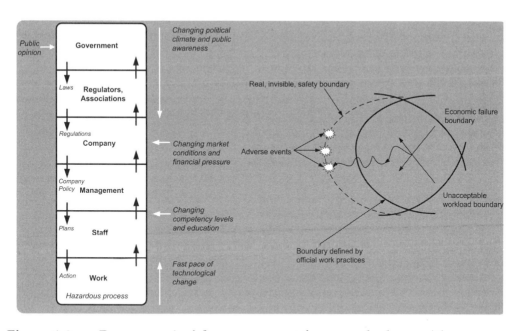

Figure 1-2   Rasmussen's risk management framework along with migration of work practices models (adapted from Rasmussen 1997)

who can either trigger accidental flows or divert normal work flows. Similar to the Swiss cheese model, the framework describes the various organisational levels (e.g. government, regulators, company, company management, staff, and work) involved in production and safety management and, in addition to the unsafe acts made by humans working on the front line, focuses on the mechanisms generating behaviour across the wider organisational system. According to the model, complex sociotechnical systems comprise a hierarchy of actors, individuals, and organisations (Cassano-Piche et al. 2009). Safety is viewed as an emergent property arising from the interactions between actors at each of the levels.

The model describes how each of the levels comprising safety critical systems is involved in safety management via the control of hazardous processes through laws, rules, and instructions. According to the framework, for systems to function safely, decisions made at the higher governmental, regulatory, and managerial levels of the system should be promulgated down and be reflected in the decisions and actions occurring at the lower levels (i.e. staff work levels). Conversely, information at the lower levels regarding the system's status needs to transfer up the hierarchy to inform the decisions and actions occurring at the higher levels (Cassano-Piche et al. 2009). Without this so-called 'vertical integration', systems can lose control of the processes that they are designed to control (Cassano-Piche et al. 2009).

According to Rasmussen (1997), accidents are typically 'waiting for release', the stage being set by the routine work practices of various actors working within the system. Normal variation in behaviour then serves to release accidents.

A second component of Rasmussen's model describes how work practices evolve over time, and in doing so often cross the boundary of safe work activities. The model presented on the right hand side of Figure 1-2 shows how economic and production pressures influence work activities in a way that, over time, leads to degradation of system defences and migration of work practices. Importantly, this migration of safe work practices is envisaged to occur at all levels of the system and not just on the front line. A failure to monitor and address this degradation of safe work practices will eventually lead to a crossing of the boundary, resulting in an unsafe event or accident.

Finally, a more recently proposed accident model worthy of mention is the increasingly popular STAMP model (Leveson 2004). Underpinned by control theory and also systemic in nature, STAMP suggests that system components have safety constraints imposed on them and that accidents are a control problem in that they occur when component failures, external disturbances, and/or inappropriate interactions between systems components are not controlled, which enables safety constraints to be violated (Leveson 2009). Leveson (2009) describes various forms of control, including managerial, organisational, physical, operational and manufacturing-based controls. Similar to Rasmussen's model described above, STAMP also emphasises how complex systems are dynamic and migrate towards accidents due to physical, social and economic pressures.

System safety and accidents are thus viewed by STAMP as a control problem. According to Leveson the following four conditions are required to enable control over complex sociotechnical systems (Leveson et al. 2003) and accidents arise when all four conditions are not achieved:

1. *Goal condition*. The controller must have a goal or goals, e.g. to maintain safety constraints (Leveson et al. 2003);

2. *Action condition.* The controller must be able to influence the system in such a way that processes continue to operate within predefined limits and/or safety constraints in the context of multiple internal and external disturbances;
3. *Model condition.* The controller must possess an accurate model of the system. Accidents typically arise from inaccurate models used by controllers;
4. *Observability condition.* The controller must be able to identify system states through feedback, which is then used to update the controller's system model.

1.3 SUMMARY

This chapter has presented a brief overview of some of the more prominent accident causation models presented in the Human Factors literature. It is clear that, as is the case with most Human Factors concepts, the notion of accident causation has evolved somewhat since the first attempts were made to describe the causal mechanisms involved. Accidents are now most commonly viewed from a systems theoretic viewpoint, as a consequence of the inadequate, inappropriate or unwanted interactions between system components occurring at and between various levels of the complex system (e.g. Leveson 2004; Reason 1990; Rasmussen 1997). This invariably renders accidents a highly complex phenomenon and places great demand on the methods used to understand them. Also notable is that a universally accepted model of accident causation is yet to emerge. Reason's model is undoubtedly the most popular; however, more recent systems theoretic approaches, such as STAMP, may be more appropriate given the complexity associated with modern complex sociotechnical systems. Despite this, STAMP has not yet received anywhere near the same amount of attention as Reason's model. Although different in many ways, the most prominent models of accident causation are in agreement on at least one thing; that is, in order to exhaustively describe accidents, the entire complex sociotechnical system should be the unit of the analysis.

2
Human Factors Methods for Accident Analysis

2.1 INTRODUCTION TO HUMAN FACTORS METHODS

Many Human Factors methods exist. For the purposes of this book, these can be categorised as follows:

1. *Data collection methods.* The starting point in any Human Factors analysis, be it for accident analysis, system design or evaluation, involves describing existing or analogous systems via the application of data collection methods (Diaper and Stanton 2004). These methods are used to gather specific data regarding a task, device, system or scenario and are critical for accident analysis efforts, since the analysis produced is heavily dependent upon the data available regarding the accident itself. Data collection methods typically used in accident analysis efforts include interviews, observation, walkthroughs, and documentation review.
2. *Task analysis methods.* Task analysis methods (Annett and Stanton 2000) are used to describe tasks and systems and typically involve describing activity in terms of the goals and physical and cognitive task steps required. They focus on 'what an operator ... is required to do, in terms of actions and/or cognitive processes to achieve a system goal' (Kirwan and Ainsworth 1992, 1). Task analysis methods are useful for accident analysis purposes since they can be used to provide accounts of tasks and systems as they should have performed or been performed (i.e. normative task/system description) and also of how the task or system actually did perform or was performed. This is useful for identifying the unsafe acts, errors, violations, and contributing factors involved.
3. *Cognitive task analysis methods.* Cognitive Task Analysis (CTA) methods (Schraagen et al. 2000) focus on the cognitive aspects of task performance and are used for 'identifying the cognitive skills, or mental demands, needed to perform a task proficiently' (Militello and Hutton 2000, 90) and describing the knowledge, thought processes and goal structures underlying task performance (Schraagen et al. 2000). CTA methods are useful for accident analysis purposes since they can be used to gather information from those involved regarding the cognitive processes involved and the factors shaping decision-making and task performance prior to, and during, accident scenarios.

4. *Human error identification/analysis methods.* In the safety critical domains, a high proportion (often over 70 per cent) of accidents are attributed to 'human error'. Human error identification methods (Kirwan 1992a, b, 1998a, b) use taxonomies of human error modes and performance shaping factors to predict any errors that might occur during a particular task. They are based on the premise that, provided one has an understanding of the task being performed and the technology being used, one can identify the errors that are likely to arise during the man-machine interaction. Human error analysis approaches are used to retrospectively classify and describe the errors, and their causal factors, that occurred during a particular accident or incident. Both approaches are useful for accident analysis purposes. Human error identification methods can be used to identify the errors that are likely to have been made given a certain set of circumstances, whereas human error analysis methods can be used to classify the errors involved in accident scenarios.
5. *Situation awareness measures.* Situation awareness refers to an individual's, team's or system's awareness of 'what is going on' during task performance (Endsley, 1995). Situation awareness measures are used to measure and/or model individual, team, or system situation awareness during task performance (Salmon et al. 2009). Situation awareness modelling approaches are useful for accident analysis purposes since they can be used to model what situation awareness was held by the system, teams and individuals involved during the accident, and also what situation awareness would have been required in order to avoid the accident. This allows the identification of situation awareness failures in accident scenarios.
6. *Mental workload measures.* Mental workload represents the proportion of operator resources that are demanded by a task or series of tasks. Mental workload measures are used to determine the level of operator mental workload incurred during task performance. Measures of mental workload are less commonly used for accident analysis purposes; however, they can be used to retrospectively determine the workload levels (i.e. overload/underload) of those involved in the accident.
7. *Team performance measures.* Teamwork is formally defined by Wilson et al. (2007, 5) as 'a multidimensional, dynamic construct that refers to a set of interrelated cognitions, behaviours and attitudes that occur as team members perform a task that results in a coordinated and synchronised collective action'. Team performance measures are used to describe, analyse and represent various facets of team performance, including the knowledge, skills and attitudes underpinning team performance, team cognition, workload, situation awareness, communications, decision-making, collaboration and co-ordination. Team performance measures are useful for accident analysis purposes when scenarios involve teams of operators undertaking work. Of particular interest here are teamwork breakdowns (e.g. lack of communication or coordination).
8. *Interface evaluation methods.* A poorly designed interface can lead to unusable products, user frustration, user errors, inadequate performance and increased performance times. Interface evaluation approaches (Stanton and Young 1999) are used to assess the interface of a product or device; they aim to improve interface design by understanding or predicting user interaction with the device in question. Various aspects of an interface can be assessed, including layout, usability, colour coding, user-satisfaction and error potential. Interface evaluation methods are useful for accident analysis purposes since they can be used to evaluate interfaces that have been implicated as a causal factor in the accident. For example, poorly designed interfaces are often cited as the cause of operator errors involved in accident scenarios across the safety critical domains.

9. *Performance time modelling methods.* Performance time modelling methods are used to model the time incurred when human operators perform tasks or sequences of tasks in complex sociotechnical systems. These approaches are useful for accident analysis purposes since they can be used to determine whether an operator performed appropriately, in terms of response times, during the accident scenario.
10. *Accident analysis methods.* Of course, there exists a whole sub-set of methods developed specifically for accident analysis purposes. These methods are used to identify the causal factors involved in accidents and take on a variety of forms (see Section 2.3).

Other forms of Human Factors methods are also available, such as system and product design approaches (e.g. allocation of functions analysis, storyboarding, scenario-based design); however, since they do not fit in with the scope of this book we have not described them here.

2.2 KEY ISSUES AND CONSIDERATIONS DURING ACCIDENT ANALYSES

To structure this discussion, a simplistic and generic accident analysis procedure is presented in Figure 2-1. This overview can be used to support discussion of some of the key issues and considerations that arise during accident analysis efforts. The issues discussed all have a bearing on the accident analysis method used and also the quality and utility of the outputs derived.

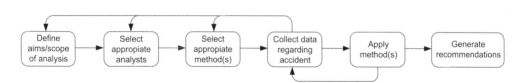

Figure 2-1   Generic accident analysis procedure

Aims and Scope

The aims and scope of the analysis are important as they have a bearing on the analysts and methods to be used, and also on the outputs generated. Although the aim of accident analysis is typically to identify the causal factors involved, often different elements of the causes or different components of the system are focused on specifically. For example, the focus may be broadly on the failures across the entire system that played a role in the accident, on the failures at one particular level of the system (e.g. government, company management) or on the failures made by human operators working on the front line. In recent times the trend is to focus on failures across the entire sociotechnical system (e.g. Cassano-Piche et al. 2009; Jenkins et al. 2010; Salmon, Williamson et al. 2010); however, different components or levels of the system are also commonly focused on, including individual operators (e.g. Stanton and Baber 2008) and equipment (Kecojevic et al. 2007).

The scope of the analysis is also important, and reflects the resources available (e.g. data, analysts, time, finances) and also the aims of the analysis. Often scope is limited by what data and analysts are available to support the analysis and how much time and financial resources are available. For example, for non-catastrophic accidents which do not involve fatalities, it is often difficult to find data regarding failures at the higher organisational levels of the system, which limits the scope to the analysis of events immediately prior to the accident. Further, time and financial constraints often mean that there is a limit to what data can be collected, and how deep the analysis can go.

Analysts and Subject Matter Experts

The analysts involved are critical to the quality of the analysis produced (Grabowski et al. 2009). It is important to select analysts who are knowledgeable in accident causation theory and Human Factors in general, and who are experienced and skilled in a range of accident analysis methods, particularly the one being used. It is also highly important to involve appropriate Subject Matter Experts (SMEs) in the analysis process, including during data collection and analysis. This is especially so for reviewing and refining analysis outputs. Often it can be useful to use a panel of analysts with different skill-sets; e.g. one Human Factors specialist, one task or domain Subject Matter Expert (SME), such as a pilot in aviation accident analyses, an engineer and one analyst with experience of the equipment involved.

Methods Used

The focus of this book – the methods used for accident analysis – obviously has a significant bearing on the output. Since various methods exist, it is important to use the analysis aims and scope and the analysts involved in identifying the chosen method. For example, for accident analyses where the failures across the entire system are of interest, a systems-based accident analysis method, such as AcciMap (Rasmussen 1997) or the Systems Theoretic Accident Modelling and Processes (STAMP; Leveson 2004) is required. Alternatively, if the aim of the analysis is to evaluate the decision-making process of one of the human operators involved in the accident, then a cognitive task analysis approach, such as the Critical Decision Method (CDM; Klein et al. 1989) should be used.

Data Available to Support Analysis

Perhaps most important is the data available to support accident analyses, with all accident analysis efforts being constrained by the data underpinning them, regardless of the efficacy of the analysis method applied (e.g. Dekker 2002; Grabowski et al. 2009; Reason 1997). For large-scale catastrophic accidents, often a great deal of data is available since government or public inquiries have invested great time and resources in investigating the accident. For the smaller accidents occurring in complex, safety critical systems, however, often the data surrounding them is sparse and does not support efficient accident analysis. For example, when applying systems-based accident analysis approaches, often sufficient data is not available to enable identification of the failures at the higher organisational or governmental levels of the system in question. This was the case in a recent analysis of Victorian (jurisdiction in Australia) fatal road traffic crashes undertaken by the authors, where it was found that the data available on road traffic crashes was only sufficient to support identification of a limited range of failures at the unsafe acts and preconditions for unsafe acts levels of the HFACS accident analysis method (Salmon, Lenné and Stephan 2010). In short, the data was found to be driver-centric and lacked sufficient detail regarding failures across the system which may have been involved, such as poor roadway/infrastructure design, poor operating procedures, and conditions promoting violations (Wagenaar and Reason 1990). This is represented in Figure 2-2, where the data that was available for our analysis is mapped onto Reason's and Rasmussen's accident causation models. Both show that the data available covered mainly the road user (e.g. driver, motorcyclist) involved and a limited sub-set of environmental (e.g. speed zone, weather conditions), equipment (vehicle type), and contextual factors (e.g. day, week, season). Clearly, a complete systems analysis is not currently supported in this jurisdiction, with higher level factors currently not reported in road traffic crash reports.

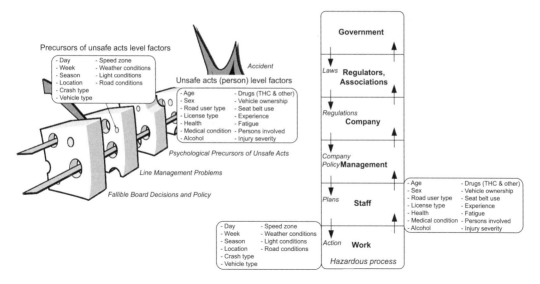

Figure 2-2   Victorian fatal road traffic crash data mapped onto accident causation frameworks

Data Collection

Given the importance of the data available regarding the accident(s) under analysis, the data collection activities employed are critical. Typical data collection activities include interviewing those involved or SMEs for the domain in question, visiting accident sites or similar work places, documentation review (e.g. existing analyses of the accident, government inquiry reports, task analyses, training manuals, procedures), observation of accident footage, observing a similar task or system, conducting walkthroughs of tasks, reconstructing events, and extracting data from accident databases, to name only a few. It is important that the data collection process is as exhaustive as possible given the constraints of the analysis, and that as much accurate data is collected as is possible.

2.3 HUMAN FACTORS METHODS FOR ACCIDENT ANALYSIS

This section presents detailed information and guidance on ten Human Factors methods that have previously been used by the authors for accident analysis purposes. Importantly, the methods consider different facets of accident scenarios, with some focusing on failures across the overall system, and others focusing on specific elements, such as human operator decision-making and task performance times. An overview of the methods described is given below and in Table 2-1.

The Critical Decision Method (CDM; Klein et al. 1989) is a cognitive task analysis method that uses semi-structured interview to gather information regarding the cognitive processes employed by human operators during critical incidents. Pre-defined cognitive probes are used to investigate decision-making at key decision points in the incident under analysis. The output is useful for accident analysis purposes since it specifies the factors influencing decision-making and performance during each key decision point.

Based on Rasmussen's risk management framework described earlier, AcciMap (Rasmussen 1997) is a systems-based accident analysis method which is used to graphically depict the decisions, actions and failures which contributed to the accident under analysis, along with the relationships between them. AcciMap considers failures by operators on the front line and also failures at the higher company management, local authority and governmental levels.

Another graphical method, Fault Tree Analysis (FTA) is used to graphically depict the failures involved in accident scenarios. Using tree-like diagrams, failure events are described along with their human - and hardware - related causes.

Inspired by Reason's Swiss cheese model, the Human Factors Analysis and Classification System (HFACS; Wiegmann and Shappell 2003) is a systems-based accident analysis method, originally developed for aviation accident analysis purposes but since applied in a range of domains, that provides analysts with taxonomies of failure modes across four organisational levels (unsafe acts; preconditions for unsafe acts; unsafe supervision; and organisational influences). Working backward from the immediate causal factors, analysts classify the errors and associated causal factors involved across the four levels using the taxonomies provided.

Viewing accidents as a control problem, Leveson's STAMP model (Leveson 2004) focuses on control failures across the overall work system. STAMP considers various forms of control, including managerial, organisational, physical, operational and manufacturing-based controls and provides analysts with a taxonomy of control failures (Leveson 2009). STAMP also uses systems dynamics to investigate the causes of the control failures identified.

Social Network Analysis (SNA; Driskell and Mullen 2004) is used to understand network structures via description, visualisation and statistical modelling (Van Duijn and Vermunt 2006). When used for accident analysis purposes SNA is used to evaluate the communications failures involved or problems within the social network (e.g. overloaded nodes or communications bottlenecks). SNA produces social network diagrams which

Table 2-1   Human factors accident analysis methods summary table

Name	Domain	Method type	Training time	App time	Input methods	Tools needed	Main advantages	Main disadvantages	Outputs
AcciMap (Svedung and Rasmussen 2002)	Generic	Accident analysis	Low	High	Interviews; Observational study; Documentation review.	Pen and paper; Flipchart; Microsoft Visio.	1. Considers both the error/failures at the sharp end and also the system-wide failures involved; 2. The output is visual and easily interpreted; 3. Considers contributory factors at six different levels, including outside of the organisation involved.	1. Time consuming due to its comprehensiveness; 2. Suffers from problems with analyst hindsight; 3. The quality of the analysis produced is entirely dependent upon the quality of the input data.	Graphical representation of the incident in question including failures at the sharp end and the causal factors across six organisational levels.
Critical Decision Method (CDM; Klein et al. 1989)	Generic	Cognitive task analysis	High	High	Interviews; Observational study.	Pen and paper; Audio recording device; Word processing software.	1. Can be used to elicit information regarding the cognitive processes involved in accident scenarios; 2. Useful for gathering data on the factors influencing decision-making during accident scenarios; 3. Has a long history of application in complex sociotechnical systems.	1. Quality of data obtained is highly dependent upon the skill of the interviewer; 2. Extent to which verbal interview responses reflect exactly the cognitive processes employed by decision makers during task performance is questionable; 3. Highly time consuming.	Transcript of cognitive processes involved in task performance along with the factors shaping decision-making.
Critical Path Analysis (CPA; Baber 2004)	Human Computer Interaction	Performance time modelling	Low	Med	Observational study; Documentation review; Hierarchical task analysis.	Microsoft Visio.	1. Uses standard operator task performance times to determine appropriate task duration times; 2. Can be useful in accident analysis efforts for determining appropriate task performance times; 3. Easy to learn and use.	1. Does not say anything about the causal factors involved in accidents; 2. Only models error free performance; 3. With the requirement for an initial task analysis, can be time consuming and difficult to apply for larger complex tasks.	Prediction of performance time associated with task sequences.

Table 2-1 Continued

Method	Domain	Application	Training time	Related methods	Tools needed	Advantages	Disadvantages	Outputs
Event Analysis of Systemic Teamwork (EAST; Stanton et al. 2005)	Generic	Teamwork assessment; Situation awareness assessment; Communications analysis; Task analysis; Cognitive task analysis.	High	High	Observational study; Interview; Documentation review; Questionnaire.	Pen and paper; WESTT software tool; HTA software tool.	1. Highly comprehensive, and activities are analysed from various perspectives; 2. The analysis produced provides compelling views of collaborative activities; 3. A number of Human Factors concepts are examined, including situation awareness, decision-making, teamwork, and communications.	1. When undertaken in full, the EAST framework is a highly time consuming approach; 2. The use of various methods ensures that the framework incurs a high training time; 3. A high level of access to the domain, task and SMEs is required.	Description of tasks performed; Rating of coordination levels; Analysis of communications; Description of situation awareness during task performance; Analysis of communications technologies; Description of cognitive processes.
Fault tree analysis (Kirwan and Ainsworth 1992)	Generic	Accident analysis	Low	Med	Interviews; Observational study; Documentation review.	Pen and paper; Flipchart; Microsoft Visio.	1. Define possible failure events and associated causes. This is especially useful when looking at failure events with multiple causes; 2. Quick to learn and use in most cases; 3. When completed correctly they are potentially very comprehensive.	1. For complex scenarios the method can be complex, difficult and time consuming to construct, and the output may become unwieldy; 2. Offers no remedial measures or countermeasures; 3. Little evidence of their application outside of the process control domains.	Graphical representation of the incident in question including failures at the sharp end and their causal factors.
The Human Factors Analysis and Classification System (HFACS; Wiegmann and Shappell 2003)	Aviation	Accident analysis	Low	High	Interview; Documentation review; Accident database.	HFACS taxonomies; Microsoft Word; Microsoft Visio; SPSS.	1. Provides taxonomies of failure modes across four levels; 2. Despite being developed specifically for aviation has been applied across a range of domains for accident analysis purposes and regularly achieves acceptable levels of reliability; 3. Sound underpinning theory.	1. For domains other than aviation the taxonomies can be limited, restricting analysts somewhat; 2. Does not consider failures outside of the organisation involved; 3. Highly dependent on the quality of the data available; e.g. it is often difficult to identify failures at the higher levels of the system involved.	Description of failures involved in accident across four levels; Statistical analysis of the relationships between failures across levels.

Table 2-1 Concluded

Method	Domain	Description		Training time	Data required	Tools needed	Advantages	Disadvantages	Outputs
TRACEr	Air Traffic Control	Error prediction and analysis technique.	Med	High	Documentation review; Observational study; Interviews.	TRACEr taxonomies	1. Provide comprehensive taxonomies covering various facets of the errors involved; 2. Can be used both predictively and retrospectively; 3. Has been applied outside of the air traffic control domain.	1. Heavily focused on the errors made by individual operators, failing to consider systems failures; 2. Can be time consuming; 3. Some of the taxonomies are specific to air traffic control.	Comprehensive analysis of operator errors involved in accidents.
Propositional networks (Salmon et al. 2009)	Generic	Model of systems awareness during task performance.	Low	High	Observational study; Verbal protocol analysis; Hierarchical task analysis.	Pen and paper; Microsoft Visio; WESTT.	1. Depict the information elements underlying the systems situation awareness and the relationships between them; 2. Can be decomposed to depict the awareness of individuals and sub-teams working within the system; 3. Avoids most of the flaws associated with measuring/modelling SA.	1. More of a modelling approach than a measure; 2. Can be highly time consuming and laborious; 3. For large scenarios involving multiple actors, the networks can become overly difficult to construct, large and unwieldy, and can be difficult to present.	Networks depicting the situation awareness held by system, teams and individual agents (human and non-human).
Social Network Analysis (SNA; Driskell and Mullen 2004)	Generic	Analysis of communications between components of system (e.g. between individuals, between teams, between humans and technological agents).	Low	High	Observational study; Documentation review; Hierarchical task analysis.	Pen and paper; Agna SNA software; Microsoft Visio.	1. Useful for identifying communications failures involved in accidents; 2. Social network diagrams provide a powerful way of representing the communications taking place during accident scenarios; 3. Networks can be analysed mathematically.	1. Data collection procedure can be overly time consuming; 2. For large, complex networks the data analysis procedure is highly time consuming; 3. For complex collaborative tasks SNA outputs can become complex and unwieldy.	Graphical representation of communications between components of system.
Systems Theoretic Accident Modelling and Processes (STAMP; Leveson 2004)	Generic	Theoretical model of accident causation.	High	High	Documentation review; Interviews; Observational study.	Drawing package (e.g. Microsoft Visio).	1. Sound underpinning theory; 2. Considers loss of control across the overall sociotechnical system; 3. Has been applied for accident analysis in a range of domains.	1. There is limited guidance available to support analysts in conducting STAMP analyses; 2. The method is more complex than others, considering constraints, control structure, structural dynamics and behavioural dynamics; 3. Reliability of the method is questionable.	Hierarchical control structure diagram; Description of failures at each level of control; Systems dynamics model of accident.

depict the connections between entities during scenarios of interest. Statistical modelling is then used to analyse the networks mathematically in order to quantify aspects of interest.

The propositional network methodology was originally developed for modelling situation awareness in complex sociotechnical systems (Salmon et al. 2009) and has since been used to model the situation awareness-related failures involved in accidents (e.g. Griffin et al. 2010). Networks representing situation awareness are constructed, and situation awareness failures (e.g. lack of awareness, erroneous awareness, failure to communicate key information) are identified through network analysis procedures.

Critical Path Analysis (CPA; Baber 2004) is used to model the performance time associated with sequences of tasks and has previously been used to model operator response times during accident scenarios (e.g. Stanton and Baber 2008). CPA works by modelling task sequences and calculating, using standard human response time data, the time required for completion of different sequences of tasks.

The Technique for the Retrospective and Predictive Analysis of Cognitive Errors (TRACEr; Shorrock and Kirwan 2002) is both an error prediction and retrospective error analysis method that has previously been used for the analysis of air traffic control and rail accidents. TRACEr uses six error taxonomies to investigate the operator errors involved, including the errors themselves, performance shaping factors, and error detection and recovery strategies.

Unlike the individual methods described above, the Event Analysis of Systemic Teamwork framework (EAST; Stanton et al. 2005) is an integrated suite of methods that is used to analyse activity in collaborative systems. EAST uses a combination of methods to describe the task, social and knowledge networks underlying collaborative activity, including task analysis, social network analysis, network analysis and teamwork assessment methods. Recent applications (e.g. Rafferty et al, In Press) have involved using the approach to model accident scenarios.

2.3.1 Critical Decision Method

Background and applications The Critical Decision Method (CDM; Klein et al. 1989) is a semi-structured interview approach that is used to gather information regarding the cognitive processes underlying decision-making during events occurring in complex sociotechnical systems. Typically, 'critical' scenarios (i.e. non-routine or accident events) are decomposed into key decision points and a series of 'cognitive probes' (targeted interview probes focusing on cognition and decision-making) are used by the interviewer to interrogate the interviewee regarding the cognitive processes underlying their performance and decision- making at each decision point. This allows detailed information to be gathered both on the decisions made and also on the factors influencing decision-making. So long as the interviewee was involved in the incident of interest, the CDM approach can be applied for a range of purposes, including accident analysis and investigation. CDM is particularly useful when the accident analysis effort is focused on exploring human operator decision-making prior to and during the accident. For example, pertinent lines of inquiry include why a human operator made a particular decision and what factors (both individual and system wide) influenced the operator's decision-making.

Domain of application The procedure is generic and can be applied in any domain; since its inception it has been applied across a range of domains, including the emergency services (Blandford and Wong 2004), aviation (Paletz et al. 2009) air traffic control (Walker et al. 2010), the military (Salmon et al. 2009), energy distribution (Salmon et al. 2008), road transport (Stanton et al. 2007), rail transport (Walker et al. 2006) and white water rafting (O'Hare et al. 2000).

Accident analysis/investigation applications CDM was recently used by the authors for investigating the factors influencing worker decision-making prior to their involvement in injury-causing incidents in the retail sector. The method is also listed as an accident analysis and investigation approach on the US Federal Aviation Administration website.

Procedure and advice Step 1: clearly define aims of the analysis. Always important in any Human Factors analysis effort, the aims of the analysis should first be clearly defined. This is important for CDM-based analyses as the analysis aims define the scenarios to be focused on, the participants to be involved, and also, most importantly, the CDM probes to be used during the interviews. Exactly what the aims of the analysis to be undertaken are should therefore first be clearly defined.

Step 2: identify scenarios to be analysed. Once the aims of the analysis are clearly defined, it is next important to identify what scenarios should be analysed. Most CDM analyses focus on non-routine events or critical incidents; however, for accident analysis purposes, specific accident scenarios are of obvious interest. Selection of the scenarios to be focused on also generally identifies the participants who should be involved. In this case, the interviewee should have had a key decision-making role at some point in the accident scenario under analysis. CDM analyses can, however, also be based on direct observation of scenarios of interest; in this case, critical incidents are identified post scenario completion by asking participants to identify any critical incidents that occurred during task performance.

Step 3: select/develop appropriate CDM interview probes. The CDM works by probing participants using pre-defined 'cognitive probes' which are designed specifically to elicit information regarding the cognitive processes undertaken during task performance. It is therefore highly important, particularly when using the method for accident analysis purposes that an appropriate set of probes are selected or developed prior to the analysis. There are a number of sets of CDM probes available in the literature (e.g. Crandall et al. 2006; Klein and Armstrong 2004; O'Hare et al. 2000); however, it may also be appropriate to develop a new set of probes depending on the analysis requirements. For example, based on a synthesis of relevant probes from the literature, the authors used the following probes (see Table 2-2) when evaluating retail store worker decision-making prior to involvement in injury-causing accidents.

Step 4: select appropriate participant(s). Once the aims, task and probes are defined, appropriate participants should be selected; again this is entirely dependent upon the analysis context and requirements. It is recommended that the participants selected were the primary decision makers in the accident scenario under analysis. Alternatively, if this is not possible then SMEs for the domain and task in question could be used.

Step 5: gather description of incident, including timeline and critical events. The CDM is normally applied based on retrospective recollection of past incidents or direct observation of the task or scenario under analysis. For accident analysis purposes it is highly likely that the analysis will be based on recollection of events. In this case the interviewer and interviewee should work together to develop a detailed description of the accident scenario. The interviewer and interviewee should develop a task model or event timeline for the scenario, which is then used to define a series (normally four or five) of critical incidents or incident phases for further analysis. These normally represent critical decision points for the scenario in question; however, distinct incident phases have also been used in the past.

Step 6: conduct CDM interviews. CDM interviews should then be conducted with the participant(s) for each critical incident or key decision point identified during Step 5. Initially the interviewee is asked to describe, in as much detail as possible, the task, incident phase or decision point being analysed. Following this, the interviewer uses the cognitive probes in a semi-structured interview format in order to elicit information regarding the cognitive processes employed by the decision maker during the task, incident phase or decision point. It is important to note that the interviewer does not limit themselves to the CDM probes; rather, the approach is flexible in that further open, closed, and probing questions can be asked depending on the analysis requirements. An interview transcript should be recorded by the interviewer and the interview should be recorded using video and/or audio recording equipment. It is normally recommended that two interviewers be used, one for driving the interview discussion and one for note-taking purposes; however, it is possible to undertake CDM interviews with only one interviewer.

Step 7: transcribe interview data. Once the interviews are completed, the data should be transcribed using a word processing software package such as Microsoft Word. It is normally useful for data representation purposes to produce CDM tables, containing the cognitive probes and interviewee responses, for each participant.

Step 8: analyse data as required. The CDM data should then be analysed accordingly based on the analysis requirements. CDM data is useful in that it can be analysed both qualitatively and quantitatively. For example, when using multiple participants, it is often useful to undertake content analyses on the responses and use frequency counts to identify key themes. Alternatively, if using the approach to focus on one particular accident, judgements on the causation factors involved can be made.

Table 2-2   Accident analysis critical decision method probes

Goal specification	What were you aiming to accomplish through this activity?
Assessment	Suppose you were to describe the situation at this point to someone else. How would you summarise the situation?
Cue Identification	What features were you looking for when you formulated your decision? How did you know that you needed to make the decision? How did you know when to make the decision?
Expectancy	Were you expecting to make this sort of decision during the course of the event? Describe how this affected your decision-making process.
Options	What courses of action were available to you? Were there any other alternatives available to you other than the decision you made? How/why was the chosen option selected? Why were the other options rejected? Was there a rule that you were following at this point?
Influencing factors	What factors influenced your decision-making at this point? What was the most influential factor that influenced your decision-making at this point?
Situation Awareness	What information did you have available to you at the time of the decision?

Table 2-2   *Concluded*

Situation Assessment	Did you use all of the information available to you when formulating the decision? Was there any additional information that you might have used to assist in the formulation of the decision?
Information integration	What was the most important piece of information that you used to formulate the decision?
Experience	What specific training or experience was necessary or helpful in making this decision? Do you think further training is required to support decision-making for this task?
Mental models	Did you imagine the possible consequences of this action? Did you create some sort of picture in your head? Did you imagine the events and how they would unfold?
Decision-making	How much time pressure was involved in making the decision? How long did it actually take to make this decision?
Conceptual	Are there any situations in which your decision would have turned out differently?
Guidance	Did you seek any guidance at this point in the task/incident? Was guidance available?
Basis of choice	Do you think that you could develop a rule, based on your experience, which could assist another person to make the same decision successfully?
Analogy/generalisation	Were you, at any time, reminded of previous experiences in which a similar/different decision was made?
Interventions	What interventions do you think would prevent inappropriate decisions being made during similar incidents in the future?

Advantages:

1. CDM can be used to elicit detailed information regarding the cognitive processes involved in accident scenarios, including the decisions faced, the factors influencing decision-making and the information used;
2. Various sets of CDM probes are available (e.g. Crandall et al. 2006; Klein and Armstrong 2004; O'Hare et al. 2000);
3. The CDM approach is particularly suited to evaluating decision-making processes and can be applied for this purpose in the context of the decisions made prior to, and during, accident scenarios;
4. The method is a popular one and has been applied in a number of different domains for a range of purposes;
5. CDM offers a good return in terms of data collected in relation to time invested;
6. Its flexibility allows all manner of Human Factors concepts to be studied, including decision-making, situation awareness, human error, workload and distraction;
7. Since it uses interviews, CDM offers a high degree of control over the data collection process. Targeted interview probes can be designed a priori and interviewers can direct interviews as they see fit; and
8. The data obtained can be treated both qualitatively and quantitatively.

Disadvantages:

1. The quality of the data obtained is highly dependent upon the skill of the interviewer and the quality and willingness to participate of the interviewee. For example, interviewees may be guarded with responses for fear of reprisals;
2. Participants may find it difficult to verbalise the cognitive components of accident scenarios, and the extent to which verbal interview responses reflect exactly the cognitive processes employed during task performance is often questionable;
3. Designing, conducting, transcribing and analysing interviews is a time consuming process which may limit the number of participants that can be used;
4. The reliability and validity of interview methods is both questionable and difficult to address. Klein and Armstrong (2004), e.g. point out that methods that analyse retrospective incidents may have limited reliability due to factors such as memory degradation;
5. A high level of expertise is required in order to use CDM to its maximum effect (Klein and Armstrong 2004);
6. Since CDM typically focuses on individual operators it could promote a blame culture;
7. CDM is susceptible to a range of interviewer and interviewee biases, including social desirability bias; and
8. It is often difficult to gain the levels of access to SMEs that are required for successful completion of CDM interviews.

Related methods   CDM is based on Flanagan's Critical Incident Technique (Flanagan 1954) and is an interview-based approach. CDM applications also often utilise observational study methods and timeline analysis methods during the scenario observation and description phases. A team CDM approach, the team Decision Requirements eXercise (DRX; Klinger and Hahn 2004), is also available.

Approximate training and application times   Although the time taken for analysts to understand the CDM procedure is minimal, the training time can be high due to the requirement for experience in interviews and for trainees to grasp cognitive psychology (Klein and Armstrong 2004). In addition, once trained in the method, analysts require significant practice until they become proficient in its application. The application time is dependent upon the probes used and the number of participants involved; however, due to the high levels of data generated and the requirement to transcribe the interview data, the overall application time is typically high. CDM interviews can take anywhere between 45 minutes and two hours to complete and the transcription process for a one to two hour CDM interview normally takes between two and four hours.

Reliability and validity   The reliability of the CDM approach is questionable. Klein and Armstrong (2004) suggest that there are concerns over reliability due to factors such as memory degradation. The validity of the approach may also be questionable, due to the difficulties associated with the verbalisation of cognitive processes. Interview approaches also suffer from various forms of bias that affect reliability and validity levels, including social desirability bias.

Tools needed At a simplistic level, CDM can be conducted using only pen and paper; however, it is recommended that the interviews are recorded using video and audio recording devices. The interviewer also requires a printed set of CDM probes for conducting the CDM interviews.

Example output CDM was recently used by the authors to investigate the factors influencing retail store worker decision-making prior to involvement in injury-causing accidents. A total of 49 CDM interviews were conducted with workers who had previously been involved in an accident as reported in the retail company's accident and injury data (see Chapter 5 for full analysis). One analyst conducted the interviews, which involved the injured store workers and focused on the decision to engage in the activity that led to the injury causing accident. An extract of one of the CDM interview transcripts is presented in Table 2-3.

Table 2-3   Retail store worker injury accident CDM transcript

STORE	XXXXXX
ACTIVITY	Lifting large product (flatpack) onto trolley with customer.
INCIDENT	Basically a heavy box was involved, we're talking a flatpack, it was a large product which we no longer stock but at the time they were around the 50–60 kilo mark. There were two customers and myself, one of the customers had a trolley, the other customer was helping me lift it onto the trolley and his mate with the trolley was being a bit of an idiot and every time we'd go to put it down he'd pull away the trolley as a bit of a joke. What actually happened is the other guy who was actually physically helping me lift this thing up, he thinks that I'm closer to the trolley and he just lets go. Basically it's nowhere near the trolley and it's landed on top of me and I've gone to get up with this heavy implement on top of me and as I've come up I've twisted a little bit and its twinged my back, so yeah, it's a case of customers being stupid and not quite doing the right thing … erm … it was a busy, busy day on a Saturday … erm … in hindsight, yes, it would have been a good idea to have another staff member to help me put it into the truck or whatever but you know, if you've got someone who is bigger and stronger than you offering to give you hand that was the thing at the time. I fell to the ground and this thing landed on top of me.
Goal specification	**What were your specific goals at this point in time? What were you aiming to accomplish through this activity?** *Help the customer get this product through the register and pay for it and get him on his way.*
Cue Identification	**What features were you looking for when you formulated your decision?** *On the day in particular we were very short staffed, extremely short staffed, and I made a wrong decision at the time, I suppose, in hindsight. I basically just, you know, the customer offered to help lift it up onto the trolley. Everything was fine when he was holding onto it and it only went pear shaped when he decided to drop it and couldn't be blowed hanging onto it any longer. The customer was another six inches taller than me and muscles on muscles, he probably looked like he could have lifted the thing by himself.*
Expectancy	**Were you expecting to make this sort of decision during the course of the event? Describe how this affected your decision-making process.** *– Well look, a lot of training has gone on since this has happened. The accident happened over two years ago … erm … basically these days I don't rely on customers at all, cos they've proven to be a backwards step and now I don't rely on customers these days.*

Table 2-3 Concluded

Options	**What courses of action were available to you? Were there any other alternatives available to you other than the decision you made?** *Get a staff member, have a hostile customer who was in a hurry to get going with about another six or seven people waiting around for my services to help them where they were, so yeah, the pressure was on that day. Trying to get another staff member at the time because we were short staffed, we had a couple of sick leaves off, erm ... I would have been waiting for quite a substantial time which would have been probably beyond the realms of how long the customer was prepared to wait.* **How/why was the chosen option selected?** *– Short staffed.* *– Customer pressure.*
Influencing factors	**What factors influenced your decision-making at this point?** *– Short staffed, I knew how busy it was, and customers were waiting.* **What was the most influential factor that influenced your decision-making at this point?** *– Customer was a little bit hostile, almost stand over tactics, 'Come on, mate, I haven't time to stuff around', effectively was his words he used, cos I did offer originally to get someone to help me lift it into the trolley and lift it onto their truck for them but he said, 'Come on, I can help you lift it', so ...*

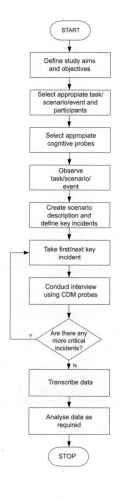

Flowchart 2-1   CDM

Recommended Reading

Crandall, B., Klein, G. and Hoffman, R. (2006) *Working Minds: A Practitioner's Guide to Cognitive Task Analysis*. Cambridge, MA.: MIT Press.

Klein, G. and Armstrong, A.A. (2004) Critical decision method. In: N.A. Stanton, A. Hedge, E. Salas, H. Hendrick and K. Brookhaus (eds), *Handbook of Human Factors and Ergonomics Methods*. Boca Raton, Florida: CRC Press, pp. 35.1–35.8.

Klein, G., Calderwood, R. and McGregor, D. (1989) Critical decision method for eliciting knowledge. *IEEE Transactions on Systems, Man and Cybernetics*, 19:3, 462–72.

2.3.2 AcciMap

Background and applications AcciMap (Rasmussen 1997; Svedung and Rasmussen 2002) is an accident analysis method that is used to represent the system-wide causal failures involved in accidents graphically. As well as identifying the environmental conditions and physical processes involved, it also focuses on the causal flow of events upstream from the accident, looking at the planning, management and regulatory bodies that may have contributed in some way (Svedung and Rasmussen 2002). Typically, the following six levels of complex sociotechnical systems are considered (although these can be modified to suit analysis needs): government policy and budgeting; regulatory bodies and associations; local area government planning and budgeting (including company management, technical and operational management; physical processes and actor activities; and equipment and surroundings). Failures at each of the levels are identified and linked between and across levels based on cause-effect relations.

Domain of application AcciMap is a generic approach that was developed for accident analysis purposes in any complex sociotechnical system. The method has been applied to a range of accidents, including gas plant explosions (Hopkins 2000), police firearm accidents (Jenkins et al. 2010), loss of space vehicles (Johnson and de Almeida 2008), outdoor education (Salmon, Williamson et al. 2010) and aviation accidents (Royal Australian Air Force 2001), public health incidents (Cassano-Piche et al. 2009; Woo and Vicente 2003; Vicente and Christoffersen 2006), and road and rail accidents (Svedung and Rasmussen 2002).

Accident analysis/investigation applications Developed specifically for accident analysis purposes, AcciMap has been applied for this purpose in a range of safety critical domains (see above).

Procedure and advice Step 1: data collection. Being a retrospective approach, the AcciMap approach is dependent upon accurate and detailed data regarding the incident under analysis. The first step therefore involves collecting detailed data regarding the incident in question, with a view to identifying failures at each of the six levels specified. Data collection for AcciMap can involve a range of activities, including interviews with those involved in the incident or SMEs for the domain in question, analysing reports or inquiries into the incident, and observing recordings of the incident. Since the AcciMap method is so comprehensive, the data collection phase is typically time consuming and involves analysing numerous data sources. Key to this approach is the collection of data regarding failures that occurred before the accident in question (i.e. over long periods of time before the accident).

Step 2: identify equipment and surroundings and physical process/actor activities failures. AcciMap analyses involve identifying the failures involved in an accident across the following six organisational levels: government policy and budgeting; regulatory bodies and associations; local area government planning and budgeting (including company management, technical and operational management; physical processes and actor activities; and equipment and surroundings). These failures are then linked both within and across levels based on cause and effect relations. It is normally straightforward to identify the failures involved at the equipment and surroundings and physical process/actor activity levels, since these are normally obvious or easy to identify (e.g. lack of equipment, adverse weather, decision to violate procedures). Since the process normally requires much iteration, it is useful to begin by constructing a rough AcciMap diagram using A3 paper, a flipchart, or a large whiteboard. The first failures identified during this step should be placed at either the equipment and surroundings and physical process/actor activity level as appropriate.

Step 3: identify factors associated with failures identified during step 2. For each failure identified during Step 2, the analyst should then use the data collected during Step 1 to identify any related factors at each of the following levels: government policy and budgeting; regulatory bodies and associations; local area government planning and budgeting; physical processes and actor activities; and equipment and surroundings. For example, for the equipment and surroundings failure of 'lack of equipment', financial restrictions imposed at the company management level which prevented the purchase of new equipment would be an example of a causal factor at the other levels. The analyst should take each failure at the equipment and surroundings and physical processes and actor activities levels and identify related failures at the other four levels.

Step 4: identify failures at other levels. The process of identifying failures at the equipment and surroundings and physical processes and actor activities level and the causal factors from the other levels is normally sufficient to identify the majority of failures involved; however, it is often useful to step through the other levels to see if any failures have been missed. Next the analyst should take each level and identify any other failures involved. If any failures are identified at the other levels then the consequences and causal factors should be identified and added to the rough diagram.

Step 5: construct draft AcciMap diagram. Using the draft diagram constructed during Steps 2–4, the analyst should then construct the first draft of the AcciMap. A software drawing package such as Microsoft Visio is normally useful for this purpose (ideally a package that allows boxes to be moved and relationships retained). When constructing the first draft (and following drafts) the analyst should check that failures are placed at the appropriate level and also that the links between failures are correct.

Step 6: finalise and review AcciMap diagram. It takes multiple reiterations before the AcciMap is complete. The final step therefore involves reviewing and revising the AcciMap diagram; however, this step might occur many times before the AcciMap is complete. Specifically, the review process involves making the following checks:

- Check that all failures have been identified;
- Check that failures are placed at the appropriate level; and
- Check that the links (cause and effect) between failures are appropriate.

One useful approach is to take each failure in isolation and to first check that it is placed at the appropriate level and then to check that it is linked appropriately to the other failures depicted in the AcciMap. It is also important to use SMEs during the review process to ensure the validity of the analysis; e.g. it is common for panels of SMEs to review the AcciMap at this stage. Verification against other AcciMap analyses can also be a valuable exercise. Failures towards the top of the diagram (at the governmental and regulatory levels) are frequently consistent between incidents and domains. It is often useful to use other analyses as a prompt to check if higher-order failures have been considered in the analysis.

Advantages:

1. AcciMap offer an approach to identifying both the failures at the sharp end and also those across the entire organisational system. When undertaken properly, the entire sequence of events is exposed;
2. Simple to learn and use and has a strong theoretical underpinning;
3. AcciMap permits the identification of system failures or inadequacies, such as poor preparation, inappropriate or inadequate government policy, inadequate management, bad design, inadequate training, and inadequate equipment;
4. Offers an exhaustive description of accidents, considering contributory conditions at six different levels. The different levels analysed allow causal factors to be traced back over months and even years;
5. The output is visual and easily interpreted;
6. AcciMap is a generic approach and can be applied in any domain;
7. Has been used in various domains for accident analysis purposes; and
8. Removes the apportioning of blame to individuals and promotes the development of systematic (as opposed to individual-based) countermeasures.

Disadvantages:

1. Can be time consuming due to its comprehensiveness;
2. Suffers from problems with analyst hindsight; e.g. Dekker (2002), suggests that hindsight can potentially lead to oversimplified causality and counterfactual reasoning;
3. The quality of the analysis produced is entirely dependent upon the quality of the input data used. Accurate and comprehensive data is not always available, so much of the investigation may be based on assumptions, domain knowledge and expertise. It is also often difficult to find data on higher level failures;
4. The output does not explicitly generate remedial measures or countermeasures; these are based entirely on analyst judgement;
5. Due to an absence of taxonomies of failure types to support analysts in classifying failures, questions remain over its reliability;
6. The approach can only be used retrospectively; and
7. For complex accident scenarios, the analysis output can become large and unwieldy.

Related methods   Using AcciMap involves considerable data collection activities and might involve the use of various data collection methods such as interviews, questionnaires and/or observational study.

Approximate training and application times   AcciMap is relatively easy to learn; however, depending upon the incident under analysis it can be highly time consuming to apply, with both the data collection procedure and the analysis itself requiring substantial effort on the part of the analyst. Based on our experiences analysts should leave approximately 1–2 weeks for data collection activities, up to one week for development of the draft AcciMap, and a further week for refining the AcciMap through SME review and re-iteration.

Reliability and validity   No reliability and validity data is presented in the literature. Since the analysis is only guided by the different failure levels, it may be that the reliability of the approach is low in some cases, as different analysts may describe events differently and also may miss contributory factors. This is likely to be problematic when the accident is complex and little is known regarding the causal factors involved.

Tools needed   AcciMap can be conducted using pen and paper. Typically, a rough AcciMap is produced using pen and paper, flipchart, or whiteboard; subsequently drawing software tools such as Microsoft Visio, Microsoft PowerPoint or Adobe Illustrator are used to construct the final AcciMap.

Example output   The following AcciMap was used to represent the failures leading up to the Hillsborough football stadium disaster which occurred on 15 April 1989. On the day of the incident the Liverpool and Nottingham Forest football clubs were due to contest the Football Association Cup Semi-final at the Hillsborough football stadium in Sheffield in South Yorkshire, England. Due to severe overcrowding outside the ground just prior to the game kicking off, an attempt was made to facilitate the entry of the fans into the ground. As a result, a major crush developed inside the ground. Ninety-six fans lost their lives, mainly due to asphyxiation, and over 400 required hospital treatment (Riley and Meadows 1995). The disaster remains the UK's worst football tragedy.

In order to demonstrate the method for future sports applications, we produced an AcciMap for the Hillsborough tragedy. The AcciMap (presented in Figure 2-3) was developed based on a review of Lord Justice Taylor's Inquiry Report (Lord Justice Taylor 1990). The analysis revealed a number of failures across five of the six levels. At the physical actor and processes level, various failures were identified, including communications failures between police units inside and outside the ground, a failure (despite requests) to call off the game prior to kick off, inadequate leadership and command, and a delay in initiating a major disaster plan. Various systemic failures that contributed to the failures on the day were also identified, including failures in the police force's planning for the game (e.g. a failure to review a previous operations order and the production of an inadequate operations order), a change of command part way through the planning process, and a lack of experience of handling similar events on behalf of the new commanding officer. At the local area and government level and regulatory bodies level a number of failures also allowed the continued use of an inadequate stadium design (e.g. the division of the terraces into pens).

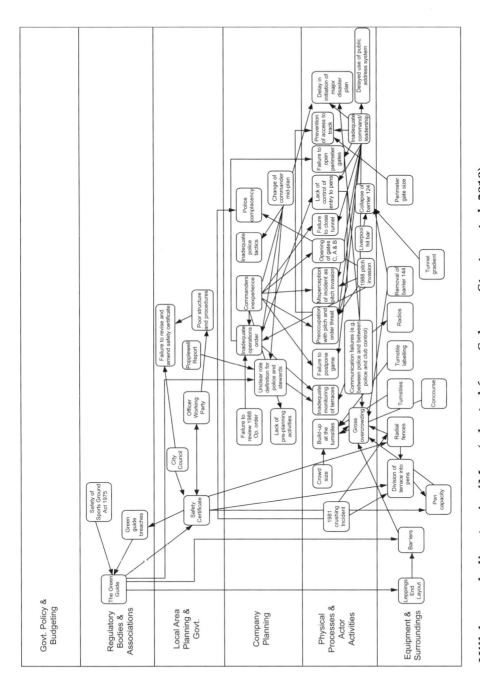

Figure 2-3   Hillsborough disaster AcciMap (adapted from Salmon, Stanton et al. 2010)

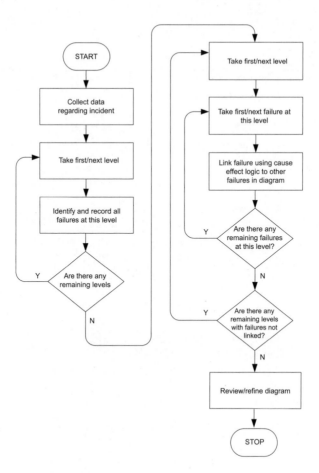

Flowchart 2-2   AcciMap

Recommended Reading

Rasmussen, J. (1997) Risk management in a dynamic society: a modelling problem. *Safety Science*, 27:2/3, 183–213.

Svedung, J. and Rasmussen, J. (2002) Graphic representation of accident scenarios: mapping system structure and the causation of accidents. *Safety Science*, 40, 397–417.

2.3.3 Fault Tree Analysis

Background and application   Fault trees are used to graphically depict the failures leading up to accidents as well as their causes. They use tree-like diagrams to define failure events and possible causes in terms of hardware failures and/or human errors (Kirwan and Ainsworth 1992). The fault tree approach was originally developed for the analysis of complex systems in the aerospace and defence industries (Kirwan and Ainsworth 1992) and is now most commonly used in probabilistic safety assessment in the process control domains. The process begins by identifying the failure or top event, which is placed at the top of the fault tree. Contributing events are then placed below using AND and OR logic gates (Kirwan and Ainsworth 1992). AND gates are used when more than one event causes a failure, whereby all the events placed directly underneath an AND gate must occur together for the failure event above to occur. OR gates, on the other hand, are used when the failure event could be caused by any one contributory event in isolation.

Domain of application Fault tree analysis was originally applied in the nuclear power and chemical processing domains. However, the method is generic and could potentially be applied in any domain.

Accident analysis/investigation applications Fault trees have been used for accident analysis purposes in a range of domains. Recent applications have taken place in the shipping (Celik et al. 2010), chemical process control (Nivolianitou et al. 2004) and hydro power plant domains (Doytchev and Szwillus 2009).

Procedure and advice Step 1: define failure event. The failure or event under analysis should be defined first. This may be either an actual event that has occurred (retrospective incident analysis) or projected failure event (predictive analysis). This event then becomes the top event in the fault tree.

Step 2: collect data regarding failure event. Fault tree analysis is dependent upon accurate data regarding the incident under analysis. The next step involves collecting data regarding the incident in question; for fault trees this can involve a range of activities, including interviews with those involved in the incident or SMEs, analysing reports or inquiries into the incident and observing recordings of the incident.

Step 3: determine causes of failure event. Once the failure event has been defined, the contributory causes associated with the event should be defined. The nature of the causes analysed is dependent upon the focus of the analysis. Typically, human error and hardware failures are considered (Kirwan and Ainsworth 1992). It is useful during this phase to use various supporting materials, such as documentation regarding the incident, task analyses outputs, and interviews with SMEs or those involved in the incident.

Step 4: AND/OR classification. Once the cause(s) of the failure event are defined, the analysis proceeds with the AND or OR causal classification phase. Each causal factor identified during Step 3 should be classified as either an AND or an OR event. If two or more contributory events combine to contribute to the failure event, then they are classified as AND events. If two or more contributory events can cause the failure even when they occur separately, then they are classified as OR events. Again, it is useful to use SMEs or the people involved in the incident under analysis during this phase.

Steps 3 and 4 should be repeated until each of the initial causal events and associated causes are investigated and described fully.

Step 5: construct fault tree diagram. Once all events and their causes have been defined fully, they should be put into the fault tree diagram. The fault tree should begin with the main failure or top event at the top of the diagram with its associated causes linked underneath as AND/OR events. Next, the causes of these events should be linked underneath as AND/OR events. The diagram should continue until all events and causes are exhausted fully or until the diagram satisfies its purpose.

Step 6: review and refine fault tree diagram. Constructing fault trees is a highly iterative process. Once the fault tree diagram is complete, it should be reviewed and refined, preferably using SMEs or the people involved in the incident.

Advantages:

1. Fault trees are useful in that they define possible failure events and associated causes. This is especially useful when looking at failure events with multiple causes;
2. A simple approach, fault trees are quick and easy to learn and use;
3. Can be used both qualitatively and quantitatively;
4. The output is easily interpreted;
5. When completed correctly they are potentially very comprehensive;
6. Could potentially be used both predictively and retrospectively; and
7. Although most commonly used in the analysis of nuclear power plant events, the method is generic and can be applied in any domain.

Disadvantages:

1. A dated approach, given theoretical advances in accident causation;
2. When used in the analysis of complex incidents, fault trees can be complex, difficult and time consuming to construct, and the output may become unwieldy;
3. Has more of a focus on hardware failures and human errors and may struggle to represent environmental conditions and higher-level organisational failures; and
4. To utilise the method quantitatively, a high level of training may be required (Kirwan and Ainsworth 1992).

Related methods   Fault tree analysis is often used with event tree analysis (Kirwan and Ainsworth 1992). Fault trees are similar to many other charting methods, including cause–consequence charts, decision action diagrams and event trees. Data collection methods, such as interviews and observational study are also typically used during the construction of fault tree diagrams.

Approximate training and application times   A simplistic method, the training time required for the fault tree method is minimal. The application time is dependent upon the incident under analysis. For complex failure scenarios, the application time is high; however, for more simple failure events, the application time is often very low.

Reliability and validity   No data regarding the reliability and validity of the fault tree approach are presented in the literature.

Tools needed   Fault tree analysis can be conducted using pen and paper; however, it is recommended that when constructing fault tree diagrams, a drawing package such as Microsoft Visio or Adobe Illustrator is used to produce the final fault tree diagram. A number of bespoke software packages are also available for drawing fault trees and these have the advantage of automating some of the probabilistic calculations.

Example output   To demonstrate the fault tree analysis method, an analysis of the *Herald of Free Enterprise* ferry disaster is presented. The *Herald of Free Enterprise* passenger ferry capsized in shallow waters just outside Zeebrugge harbour on 6 March 1987, killing 150 passengers and 38 crew members. The immediate cause of the disaster, the flooding of the lower deck, was attributed to the ferry setting sail with its inner bow doors open, which was a result of the assistant bosun's failure to close the ship's bow doors and the

CHAPTER 2 • HUMAN FACTORS METHODS FOR ACCIDENT ANALYSIS   31

captain's decision to set sail with the bow doors open (although the captain was not aware of this). Various contributory factors have since been identified (see Reason 1990 for a full description). For example, at the time when he should have been closing the bow doors, the assistant bosun was in fact asleep in his cabin, having been relieved of his duties and suffering from fatigue. Further, the bosun noticed that the bow doors were open; however, he felt that it was not his job to shut them and so failed to do so. Other failures involved included a general pressure placed on crews to depart early, a failure by the company to install a bow door indicator on the bridge (as requested by the ship's captain prior to the incident), and the unsafe top-heavy design of the ferry involved (Reason 1990).

A fault tree diagram was constructed based on a review of various sources of information regarding the incident. These included the official Department of Transport investigation report (1987) and other accounts of the incident such as Reason (1990) and Wikipedia (2010).

Flowchart 2-3   Fault tree

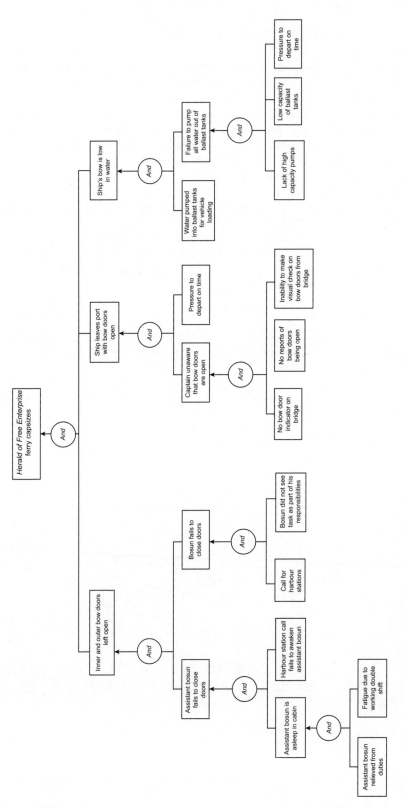

Figure 2-4 *Herald of Free Enterprise* incident fault tree extract

Recommended Reading

Kirwan, B. and Ainsworth, L.K. (1992) *A Guide to Task Analysis*. London, UK: Taylor and Francis.

2.3.4 Human Factors Analysis and Classification System

Background and applications The Human Factors Analysis and Classification System (HFACS; Wiegmann and Shappell 2003) was developed for use in the analysis of civil and military aviation accidents. The impetus came from the absence of taxonomies of latent failures and unsafe acts within Reason's Swiss cheese model, an omission which limited its application as an accident analysis framework. Accordingly, Wiegmann and Shappell (2003) set out to develop detailed taxonomies of failure modes for each level of the model to enable its application as an aviation accident analysis method. HFACS was subsequently developed based on Reason's model and an analysis of aviation accident reports (Wiegmann and Shappell 2003). It uses the following four levels, each containing different categories of failures along with their own accompanying taxonomy of failure modes: unsafe acts; preconditions for unsafe acts; unsafe supervision; and organisational influences. The structure of the HFACS method is presented in Figure 2-5, which shows the different categories mapped onto Reason's model. Working backward from the immediate causal factors, analysts classify the errors and associated causal factors involved using the HFACS taxonomies.

For demonstrative purposes, the taxonomy of errors and violations at the unsafe acts level are presented in Table 2-4. The complete taxonomies can be found in Wiegmann and Shappell (2003).

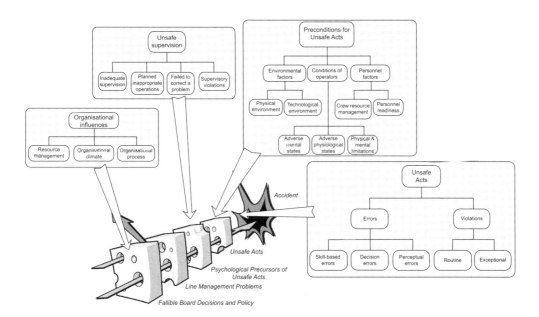

Figure 2-5 HFACS taxonomies overlaid on Reason's Swiss cheese model

Table 2-4 Unsafe acts level external error mode taxonomies (adapted from Wiegmann and Shappel, 2003)

Errors	Violations
Skill-based Errors:	*Routine*:
- Breakdown in visual scan	- Inadequate briefing for flight
- Inadvertent use of flight controls	- Failed to use ATC radar advisories
- Poor technique/airmanship	- Flew an unauthorised approach
- Over-controlled the aircraft	- Violated training rules
- Omitted checklist item	- Failed VFR in marginal weather conditions
- Omitted step in procedure	- Failed to comply with departmental manuals
- Over reliance on automation	- Violation of orders, regulations, SOPs
- Failed to prioritise attention	- Failed to inspect aircraft after in-flight caution light
- Task overload	*Exceptional*:
- Negative habit	- Performed unauthorised acrobatic manoeuvre
- Failure to see and avoid	- Improper takeoff technique
- Distraction.	- Failed to obtain valid weather brief
Decision Errors:	- Exceeded limits of aircraft
- Inappropriate manoeuvre/procedure	- Failed to complete performance computations
- Inadequate knowledge of systems and procedures	- Accepted unnecessary hazard
- Exceeded ability	- Not current/qualified for flight
- Wrong response to emergency	- Unauthorised low altitude canyon running
Perceptual Errors:	
- Due to visual illusion	
- Due to spatial disorientation/vertigo	
- Due to misjudged distance, altitude, airspeed, clearance	

Domain of application Although originally developed for the aviation domain, the flexibility and utility of HFACS is such that it has since been applied for accident analysis purposes in a wide range of safety critical domains, including general, civil and military aviation (e.g. Lenné et al. 2008; Li and Harris 2006; Li et al. 2008; Olsen and Shorrock 2010; Shappell et al. 2007), helicopter maintenance (Rashid et al. 2010), coal mining (Patterson and Shappell 2010), rail transport (Baysari et al. 2008; Reinach and Viale 2006), maritime (Celik and Cebi 2009), construction (Walker 2007), healthcare (El Bardissi et al. 2007) and road transport (Iden and Shappell 2006).

Accident analysis/investigation applications Originally conceived for accident analysis purposes, HFACS has been applied in numerous accident analysis efforts. For example, HFACS has been applied many times in the context of general, civil and military aviation accidents (e.g. Lenné et al. 2008; Li and Harris 2006; Li et al. 2008; Wiegmann and Shappell 2003)). Further, the approach has been applied for accident analysis purposes across various safety critical domains (see domain of application section above).

Procedure and advice Step 1: data collection. HFACS is dependent upon accurate data regarding the incident(s) under analysis. The first step therefore involves collecting detailed data regarding the cases to be analysed. Typically, HFACS is used to analyse multiple accident cases (e.g. a year's worth of accident data from a particular organisation), so this might involve gathering accident and incident data from accident databases or integrating data sets. Accident databases typically include a narrative of each incident along with some judgement on the causal factors involved; however, other useful data might also be available, including video recordings of the incident, investigation reports, and accounts from those involved.

Step 2: assemble analysts/panel. Often multiple analysts and/or a panel of SMEs are used for HFACS analyses. It is important before any analysis takes place to assemble the team of appropriate analysts to be involved. For analysts applying the method, researchers and practitioners with experience of the HFACS method, experience of the domain in question, and knowledge of contemporary accident causation models are preferable. It is normally useful to provide some form of training in the HFACS method to ensure that analysts have a consistent understanding of the different failure modes. If an expert panel is being used to validate the analyses, a combination of domain SMEs and personnel experienced in the HFACS method is useful.

Step 3: begin analysis by identifying unsafe acts involved. When analysing cases, the first step involves identifying the unsafe acts involved. Since HFACS uses taxonomies of external error and failure modes, this involves using the data available to classify any errors or violations that were made by front line workers (e.g. pilots, miners) that led to the accident occurring. Within the errors category the following three basic error types are defined: skill-based errors, decision errors, and perceptual errors. The violations category comprises different forms of routine and exceptional violations. It is preferable to classify specific error or violation types within one of these categories (e.g. breakdown in visual scan, exceeded ability, perceptual error due to visual illusion); however, if the data is not detailed enough then a basic classification of one of the five error/violation categories is acceptable (e.g. skill-based error).

Step 4: identify failures at the pre-conditions for unsafe acts level. Next the analyst classifies any pre-conditions for unsafe acts involved in the accident. Preconditions for unsafe acts refer to the underlying latent conditions that contribute to the occurrence of unsafe acts. This level comprises the following three categories: conditions of operators, environmental factors, and personnel factors. The conditions of operators category includes adverse mental states (e.g. distraction, mental fatigue, loss of situational awareness), adverse physiological states (impaired physiological state, medical illness, physical fatigue) and physical/mental limitations (insufficient reaction time, visual limitation, incompatible physical capability). The environmental factors category includes physical environment factors (e.g. weather, lighting) and technological environment factors (e.g. equipment/control design, automation). The personnel factors category includes crew resource management factors (e.g. lack of teamwork, failure of leadership) and personnel readiness factors (e.g. inadequate training, poor dietary practice). It is important when identifying failures at this level to link them using cause and effect logic to failures at the unsafe acts level.

Step 5: identify failures at the unsafe supervision level. The third level within HFACS, unsafe supervision, considers those instances where supervision was either lacking or inappropriate. According to Wiegmann and Shappell (2003) the role of any supervisor is to provide workers with the opportunity to succeed, and this is achieved through the provision of guidance, training, leadership, oversight, and incentives. The unsafe supervision category comprises four categories of supervisory system failures: inadequate supervision, planned inappropriate operations, failure to correct a known problem, and supervisory violations. Inadequate supervision refers to those instances when efficient supervision was not provided, with examples including 'failed to provide proper training', 'failed to provide professional guidance/oversight', and 'failed to provide adequate rest period'. Planned inappropriate operations refer to those instances where the operational tempo and/or scheduling of aircrew put individuals at unacceptable risk, jeopardised crew rest and/or affected performance (Wiegmann and Shappell 2003). Examples of planned inappropriate operations include 'poor crew pairing' and 'failed to provide adequate opportunity for crew rest'. The failure to correct a known problem category refers to those instances when a supervisor is aware of inadequacies within the system, such as inadequate equipment, training or individuals, but does not attempt to rectify them. Examples include 'failed to correct inappropriate behaviour/identify risky behaviour' and 'failed to report unsafe tendencies'. Finally, the supervisory violations category refers to those instances when rules and regulations are wilfully disregarded by supervisors (Wiegmann and Shappell 2003). Examples of supervisory violations include 'authorised unqualified crew flight' and 'violated procedures'.

Step 6: identify failures at the organisational influences level. The final level within the HFACS framework is the organisational influences level. At this point analysts are looking for failures within the higher managerial levels of the organisation that played a role in the accident. Three categories of organisational influences are used: resource management (e.g. staffing/manning, excessive cost cutting, poor design); organisational climate (e.g. structure, policies and culture); and organisational process (e.g. time pressure, instructions, risk management).

Step 7: link failures across levels. One useful aspect of HFACS analyses is the ability to link failures across the four levels. This step involves taking each failure identified and determining what causal factors exist within and across levels, and also to what other failures the failure being considered acts as a causal factor.

Step 8: review and refine analysis. It takes multiple reiterations before the HFACS analysis is complete. The final step therefore involves reviewing and revising the analysis; however, this step might occur many times before it is complete. Specifically, the review process involves making the following checks:

- Check that all failures have been identified;
- Check the appropriate HFACS failure modes have been classified; and
- Check that the links (cause and effect) between failures across and within levels are appropriate.

One useful approach is to take each failure in isolation and to first check that it is placed at the appropriate level and then to check that it is linked appropriately to the other failures. It is also important to use SMEs during the review process to ensure the validity of the analysis; e.g. it is common for panels of SMEs to review the outputs at this stage. Verification against other HFACS analyses can also be a valuable exercise.

Step 9: calculate inter-rater reliability statistics and resolve disagreements. When more than one analyst is used (as is preferable), it is accepted practice to calculate inter-rater reliability statistics. This normally involves the use of standard reliability tests such as Cohens Kappa or signal detection theory sensitivity index calculations. Also at this stage any disagreements between analysts are resolved through further discussion until consensus is reached.

Step 10: analyse outputs using frequency counts. When multiple accident cases are analysed, simple frequency counts are used initially to derive an overview of the analysis. This involves calculating the frequency with which failures at each of the four levels are involved in the accidents analysed. For example, skill-based errors are typically the most commonly identified error type at the unsafe acts level (Baysari et al. 2008; Patterson and Shappell 2010; Wiegmann and Shappell 2003).

Step 11: analyse associations between failures across the different HFACS levels. A useful aspect of the HFACS approach is the ability to analyse statistically the associations between failures across the different levels (i.e. the probability that a factor at one HFACS level predicts the presence of a factor at another level). This ensures appropriate countermeasure development (i.e. failures across the overall organisational system are dealt with as opposed to merely front line operator errors) and also allows associations to be compared across domains. One way of examining the associations between failures across the different levels is through the use of Fisher's Exact Test for contingency tables, where odds ratios (OR) are calculated to assess the strength of association. Odds are calculated for lower level factors – the odds is the ratio of the probability that a (lower-level) factor is present, to the probability that it is absent. The odds can be calculated under two conditions: one for when a higher-level factor is present, and another for when a higher-level factor is absent. An odds ratio is calculated by dividing these two odds.

Advantages:

1. HFACS offers an approach for identifying both the failures at the sharp end and also those across the entire organisational system. This promotes the development of systematic (as opposed to individual-based) countermeasures;
2. Simple to learn and the output is easily interpreted;
3. Based on Reason's widely accepted and applied Swiss cheese model of organisational accidents;
4. Provides analysts with taxonomies of failures at four different levels;
5. Allows the associations between failures at the different levels to be analysed statistically;
6. Typically achieves acceptable levels of inter-rater reliability;
7. Is a useful approach for analysing multiple accidents;
8. Although developed specifically for aviation, many of the failure modes are generic, allowing HFACS to be applied in any domain; and
9. A popular approach, HFACS has been applied in many safety-critical domains resulting in numerous peer-reviewed journal articles.

Disadvantages:

1. Unlike AcciMap, the levels considered do not go beyond the organisation involved, leaving failures outside of the organisation unexplored (e.g. legislation failures, government and local authority failures);
2. Can be time consuming due to its comprehensiveness;
3. Although enhancing reliability, the use of taxonomies of error and failure modes constrains the analyst in terms of what errors/failures can be identified. For example, when used outside of aviation some of the error/failure modes may not apply;
4. Suffers from problems with analyst hindsight; e.g. Dekker (2002), suggests that hindsight can potentially lead to oversimplified causality and counterfactual reasoning;

5. The quality of the analysis produced is entirely dependent upon the quality of the input data used. Often, e.g. data does not support classification of specific unsafe acts or higher organisational influences level failures; and
6. The output does not generate remedial measures or countermeasures; these are based entirely on analyst judgement.

Related methods HFACS is based on Reason's Swiss cheese model of organisational accidents. A number of domain specific or extended versions of the framework have been proposed, such as HFACS-ME (Rashid et al. 2010), which was proposed for the analysis of helicopter maintenance failures.

Approximate training and application times Providing the analyst has some prior knowledge of psychology and contemporary accident causation models, such as Reason's Swiss cheese model, HFACS requires very little training time. The application time incurred is dependent on the number of cases being analysed, but is typically time consuming due to the comprehensive nature of the analysis.

Reliability and validity A number of HFACS studies report inter-rater reliability statistics in terms of the level of agreement between different raters using the approach to analyse the same incident(s). Agreement ratings generally range between acceptable and high (e.g. Lenné et al. 2008; Li and Harris 2006; Li et al. 2008).

Tools needed HFACS can be applied using pen and paper, although analysts require the HFACS taxonomies to support the analysis (found in Wiegmann and Shappell 2003). Typically a standard drawing software package is used to create analysis outputs, and statistical packages, such as SPSS, are used to calculate reliability statistics and to analyse the associations between the failures identified at different levels.

Example output As part of an accident analysis methods comparison study, HFACS was recently used by the authors to analyse the Lyme Bay sea canoeing tragedy in which four students drowned whilst participating in an introductory sea canoeing activity (see Chapter 3). The incident involved a group of eight students, their schoolteacher, a junior instructor and a senior instructor participating in an introductory open sea canoeing activity in Lyme Bay, Dorset in the UK. After entering the water, the schoolteacher suffered a series of initial capsizes close to shore. Whilst the senior instructor attempted to right the school teacher after he continually capsized shortly after entering the water, the junior instructor and the eight students became separated from them and were blown out to sea. Whilst out at sea, high wind and wave conditions, lack of appropriate equipment, and inexperience led to all of the canoes being swamped and subsequently capsizing. As a result, the eight students and the junior instructor were left in the water with all canoes abandoned. After a delayed response and rescue attempt four students lost their lives through drowning. An official inquiry into the incident found various failures related to the instructors and activity centre, the company running the activity centre, and also government legislation at the time.

One analyst with significant experience in applying the HFACS method conducted an analysis of the incident using the official inquiry report as the primary data source. The AcciMap method was also used to analyse the incident. The HFACS analysis is presented in Figure 2-6.

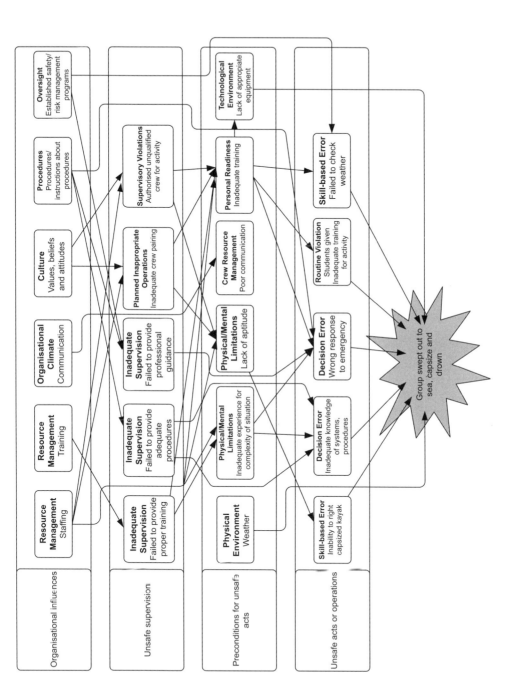

Figure 2-6   HFACS analysis of Lyme Bay sea canoeing tragedy

The analysis identified failures at all four HFACS levels. Notably, certain failures could not be classified using the standard HFACS taxonomies. For example, governmental failures (lack of legislation or regulatory body) could not be represented. In conclusion to the analysis, it was reported that, although HFACS provided a useful description of the incident, two major flaws limited its utility when compared to the AcciMap approach. First, the use of taxonomies of specific error and failure modes (developed originally for aviation accident analysis) meant that some of the errors and failures involved could not be identified. The analyst is effectively constrained by the taxonomies in terms of what contributory factors can and cannot be identified. This is problematic because often the failures identified by the analyst do not fit neatly into one of the error or failure modes provided by HFACS. As a corollary, analysts often force the failures identified to fit into one of the HFACS error or failure modes. Second, since HFACS only considers failures within the organisation in question (i.e. organisational influences is the highest level), failures outside of the organisation involved could not be considered and were thus overlooked. In the case of the Lyme Bay incident, e.g. local authority and governmental failures were implicated by the official inquiry into the incident; however, these could not be represented in the HFACS analysis since they were outside the taxonomies provided.

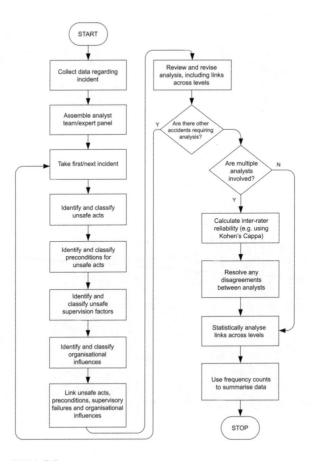

Flowchart 2-4   HFACS

Recommended Reading

Wiegmann, D.A. and Shappell, S.A. (2003) *A Human Error Approach to Aviation Accident Analysis. The Human Factors Analysis and Classification System*. Burlington, VT: Ashgate Publishing Ltd.

2.3.5 STAMP

Background and application   The STAMP accident causation model described by Leveson (2004) has also been applied as an accident analysis method in various domains (e.g. Leveson 2004; Leveson et al. 2003). STAMP focuses on hierarchical levels of control and constraints, purporting that accidents occur when the control of safety related constraints fails. According to the model, system components have safety constraints imposed on them and accidents are a control problem in that they occur when component failures, external disturbances, and/or inappropriate interactions between system components are not controlled, which enables safety constraints to be violated (Leveson 2009). Leveson (2009) describes various forms of control, including managerial, organisational, physical, operational and manufacturing-based controls. STAMP analyses therefore focus on the control structure of the system in question and the failures involved at each level of the control structure.

Domain of application   The STAMP model is generic and can be applied in any complex sociotechnical system.

Accident analysis/investigation applications   STAMP has been applied for accident analysis purposes in a range of areas, including public health (e.g. the Walkerton E.coli incident; Leveson et al. 2003), the military (Leveson et al. 2002), aviation (e.g. Nelson 2008) and air traffic management (Arnold 2009). The approach has also been used in combination with HFACS for the analysis of aviation accidents (e.g. Harris and Li 2011).

Procedure and advice   Step 1: data collection. Like all accident analysis methods, STAMP is dependent upon accurate data regarding the accident in question. The first step therefore involves collecting detailed data regarding the accident to be analysed and also importantly the domain and organisation in which the accident took place. Typically, STAMP is used to analyse single accidents in isolation (e.g. Arnold 2009; Leveson et al 2003) and focuses on the accident and also the control structure of the domain itself. Data collection is therefore likely to involve various activities, including reviewing accident reports, inquiry reports, and task analyses of the system in question; interviewing personnel involved in the accident; reviewing documents regarding the domain in question (e.g. rules and regulations, standard operating procedures); and/or interviewing SMEs for the domain/system in question.

Step 2: construct hierarchical safety control structure diagram. Once sufficient data regarding the accident and domain in question is collected, the next step involves determining the hierarchical safety control structure present for the system under analysis. This involves the construction of a hierarchical safety control structure diagram. Leveson (2004) describes how systems theory views systems as hierarchical structures where each level imposes constraints on the activities of the level beneath them. The system under analysis therefore needs to be described in terms of its hierarchical safety control structure. A generic control structure diagram is presented in Figure 2-7 (adapted from Leveson 2004). The left-hand side of the diagram shows the control structure for system development, whereas the right-hand side shows the control structure for system operations. The arrows between the levels represent communications that are used to impose constraints on the levels below, and to provide constraints on the levels below them, and to provide feedback to the levels above regarding how effective the constraints are (Leveson 2004). Although each domain/system will have its own unique control structure, it is likely to be similar in structure to that presented in Figure 2-7 (Leveson 2004).

Step 3: identify failures at each level of the control structure. According to the STAMP model, accidents result from inadequate control of safety-related constraints (Leveson 2004). The next step involves identifying failures in the control loops at each level of the control structure model produced during Step 2. Leveson (2004) proposes a taxonomy of control failures (see Table 2-5), including three main categories of control failure: inadequate control of actions; inadequate execution of control actions; and inadequate or missing feedback. Analysts should apply the taxonomy to each control loop to identify the control failures involved. It is normally useful to represent the control failures identified on the control structure diagrams developed during Step 2.

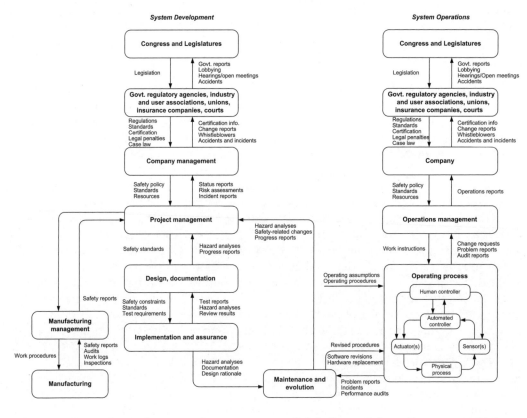

Figure 2-7 Generic control structure model (adapted from Leveson 2004)

CHAPTER 2 • HUMAN FACTORS METHODS FOR ACCIDENT ANALYSIS   43

Table 2-5   STAMP's control failure categories (adapted from Harris and Li 2011)

Inadequate enforcement of constraints:
- Unidentified hazards
- Inappropriate, ineffective or missing control actions
- Failure of control process to enforce constraints
- Inconsistent/incomplete/incorrect process models
- Inadequate coordination between controllers and decision makers

Inadequate execution of control actions:
- Communication failures
- Inadequate actuator operations
- Time lag

Feedback failures:
- Not provided through system design
- Communications failures
- Time lag
- Inadequate sensor operation

Step 4: construct systems dynamics model of accident. Although STAMP analyses can end once control failures have been identified, Leveson et al. (2003) also propose a systems dynamics approach for examining the causes of the control failures identified. Leveson et al. (2003) argue that this allows analysts to understand the processes underpinning the control structure changes which led to control failures. An extract of Leveson et al.'s (2003) systems dynamics model for the Walkerton water contamination incident is presented in Figure 2-8. Within the diagram, directional arrows are used to represent causality, and positive and negative associations are used to depict the direction of the ensuing variable change. For example, enhanced operator competence is likely to improve the effectiveness of the water quality control system.

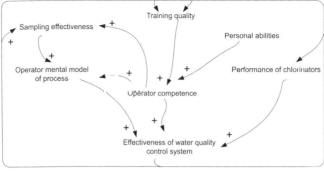

Figure 2-8   Extract of systems dynamics model of Walkerton water contamination incident (adapted from Leveson et al. 2003)

Step 5: review and finalise analysis. It takes multiple reiterations before the STAMP analysis is complete. The final step therefore involves reviewing and revising the STAMP outputs; however, this step might occur many times before the overall analysis is complete. Specifically, the review process involves making the following checks:

- Check that all control failures have been identified; and
- Check that the control failures are classified appropriately.

It is important to use SMEs during the review process to ensure the validity of the analysis; e.g. it is common for panels of SMEs to review the STAMP outputs at this stage. Verification against other STAMP analyses can also be a valuable exercise.

Advantages:

1. The analysis output is likely to be highly exhaustive, covering failures at all levels of the complex sociotechnical system. STAMP models the entire control structure and considers the contribution of designers, operators, managers and regulators to accidents (Leveson 2004);
2. STAMP is generic and can be applied for accident analysis purposes in any complex sociotechnical system;
3. The approach is based on sound underpinning theory and is consistent with the contemporary Human Factors systems approach (e.g. Hollnagel 2004);
4. Removes the apportioning of blame to individuals and promotes the development of systematic (as opposed to individual-based) countermeasures;
5. Has been applied for accident analysis purposes in various domains;
6. Provides a taxonomy of control failures to support analysts in identifying and classifying control failures; and
7. According to Leveson (2004) the approach supports the identification of countermeasures and strategies for accident prevention.

Disadvantages:

1. Likely to be highly time consuming and resource intensive in most cases. Since the control structure of the domain is described, data collection goes beyond the accident itself and requires access to SMEs;
2. Due to the lack of published guidance for analysts the reliability of the method is likely to be low;
3. Is more complex to grasp than other methods, both in terms of the underpinning theory and the methods used;
4. The quality of the analysis produced is entirely dependent upon the quality of the input data used;
5. For complex accident scenarios, the analysis output can become large and unwieldy;
6. The output does not generate remedial measures or countermeasures, these are based entirely on analyst judgement; and
7. Likely to be less useful when used for the analysis of smaller scale accidents, since the data required is often not readily available, and the effort required to determine the domains control structure will still be considerable.

Related methods The STAMP method is based on the accident causation model proposed by Leveson (2004). Using STAMP involves considerable data collection activities such as interviews, questionnaires, observational study, and documentation review.

Approximate training and application times STAMP is probably more complex than other accident analysis approaches and requires more training than most. The application time, although dependent upon the accident under analysis, is likely to be high, with both the data collection procedure and the analysis itself requiring substantial effort on the part of the analyst. Since the domain itself is described as well as the accident, the application time is likely to be higher than other approaches.

Reliability and validity No reliability and validity data is presented in the literature. Since there is limited guidance available for conducting STAMP analyses, it may be that the reliability of the approach is low in some cases, as different analysts may describe events and systems differently and also may miss control failures at each level of the control structure. This is likely to be problematic when the accident is complex and little is known regarding the causal factors involved.

Tools needed STAMP can be conducted using pen and paper. Typically, STAMP outputs are constructed in tabular and diagrammatic fashion, and so pen and paper, flipchart, or whiteboard, and then drawing software tools such as Microsoft Visio or Adobe Illustrator are used.

Example output On 15 April 2008 six students and their teacher drowned during a flash flood whilst participating in a gorge walking activity in the Mangatepopo Gorge, Tongariro National Park, New Zealand. As part of a programme of research, the overall aim of which is to develop an accident and injury surveillance system for the led outdoor activity domain in Australia, a range of accident analysis methods were applied to the analysis of this incident. The aim of these analyses was to investigate different accident analysis frameworks for their utility in describing the accidents that occurred in this domain, and also to generate evidence for the systems perspective (e.g. Rasmussen 1997; Reason 1990) as an appropriate framework for accident causation in led outdoor activities. Extracts of the STAMP analysis of this incident are presented here.

The incident occurred on 15 April 2008 when a group of ten students, their teacher, and an instructor from the Sir Edmund Hillary Outdoor Pursuit Centre (OPC), were completing a gorge walking activity in the Mangatepopo Gorge in the Tongariro National Park, New Zealand. Due to heavy rain in the area, a flash flood occurred which led to increased river flow and a rising river level in the gorge. As a result, the group eventually had to abandon the gorge walking activity and became trapped on a small ledge above the water. As the river level continued to rise, fearing the group would be washed off the ledge the instructor decided to attempt to evacuate the group from the ledge and gorge through entering the river, with poor swimmers tied to stronger swimmers, following which the instructor would extract them downstream using a 'throwbag' river rescue technique in which a bag attached to a length of rope is thrown to the person in the water and then used to pull them to the river bank. The attempted evacuation failed and only the instructor and one student managed to get out of the river as intended, with the remaining nine students and teacher being swept downstream and then over a spillway. Six students and their teacher eventually drowned. In the aftermath of the incident, the coroner and an independent investigation initiated by the activity centre involved identified various failures on behalf of the instructor and the outdoor activity centre (e.g. Brookes et al. 2009).

The basic control structure for the led outdoor activity system is presented in Figure 2-9. Figure 2-10 presents some of the control failures involved in the incident, focusing specifically on the failures in control structures between the field manager and instructor, and the instructor and group of students.

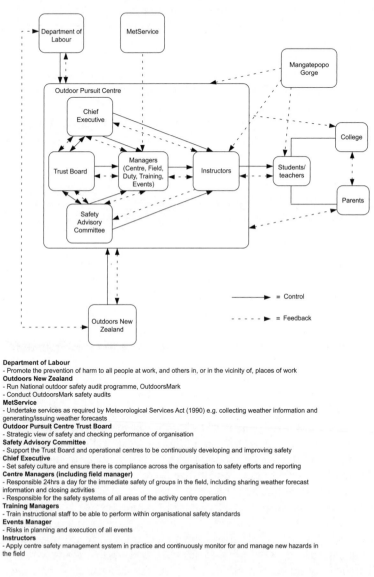

Department of Labour
- Promote the prevention of harm to all people at work, and others in, or in the vicinity of, places of work
Outdoors New Zealand
- Run National outdoor safety audit programme, OutdoorsMark
- Conduct OutdoorsMark safety audits
MetService
- Undertake services as required by Meteorological Services Act (1990) e.g. collecting weather information and generating/issuing weather forecasts
Outdoor Pursuit Centre Trust Board
- Strategic view of safety and checking performance of organisation
Safety Advisory Committee
- Support the Trust Board and operational centres to be continuously developing and improving safety
Chief Executive
- Set safety culture and ensure there is compliance across the organisation to safety efforts and reporting
Centre Managers (including field manager)
- Responsible 24hrs a day for the immediate safety of groups in the field, including sharing weather forecast information and closing activities
- Responsible for the safety systems of all areas of the activity centre operation
Training Managers
- Train instructional staff to be able to perform within organisational safety standards
Events Manager
- Risks in planning and execution of all events
Instructors
- Apply centre safety management system in practice and continuously monitor for and manage new hazards in the field

Figure 2-9   Mangatepopo incident: basic control structure diagram

CHAPTER 2 • HUMAN FACTORS METHODS FOR ACCIDENT ANALYSIS

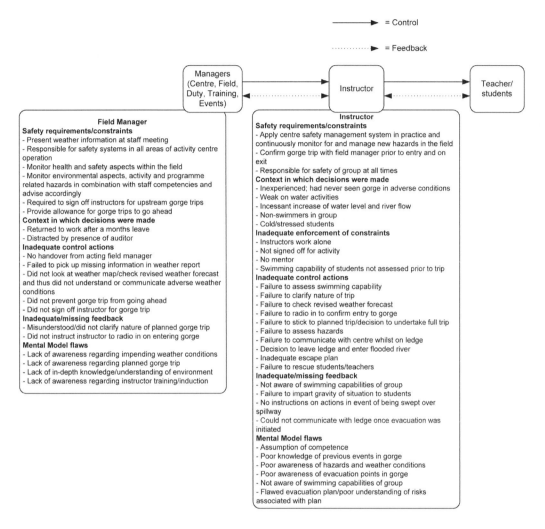

Figure 2-10 Example failures in control structures between field manager and instructor, and between instructor and students

Recommended Reading

Leveson, N.G. (2004) A new accident model for engineering safer systems. *Safety Science*, 42:4, 237–70.

Leveson, N.G., Daouk, M., Dulac, N. and Marais, K. (2003) A systems theoretic approach to safety engineering. Paper presented at Workshop on Investigation and Reporting of Incidents and Accidents (IRIA), 16 September 2003, Virginia, USA.

2.3.6 Social Network Analysis

Background and applications   Social Network Analysis (SNA; Driskell and Mullen 2004) is used to understand network structures via description, visualisation and statistical modelling (Van Duijn and Vermunt 2006). The approach is commonly applied to the analysis of relationships between agents (both human and non-human) within complex sociotechnical systems. For example, recent applications have seen SNA applied to the analysis of communications within the emergency services (e.g. Houghton et al. 2006), the military (e.g. Stanton, Salmon et al. 2010), terrorism (e.g. Skillicorn 2004) and railway maintenance (e.g. Walker et al. 2006). Importantly, the approach is not, however, limited to the analysis of communications. Houghton et al. (2006), e.g. point out that SNA is increasingly being used to investigate various phenomena, including organisation structures, financial transactions, and the spread of disease.

SNA involves collecting data, typically via observation, interview or questionnaire (Van Duijn and Vermunt 2006) regarding the relationship (e.g. communications) between the entities in the group or network under analysis. This data is then used to construct a social network diagram which depicts the connections between entities in a way in which the relationships between agents and the structure of the network can be easily ascertained (Houghton et al. 2006). Statistical modelling is then used to analyse the network mathematically in order to quantify aspects of interest, such as key communications 'hubs' or communications bottlenecks.

Domain of application   SNA is a generic approach and can be applied in any domain.

Accident analysis/investigation applications   SNA has been used as part of the EAST framework for analysing friendly fire incidents (e.g. Rafferty et al. forthcoming).

Procedure and advice   Step 1: define analysis aims. First, the aims of the analysis should be clearly defined. For example, the aim may be to evaluate the communications between specific agents, or to identify communications failures and bottlenecks during a particular accident scenario. Clear definition of the aims allows appropriate scenarios to be used and ensures that relevant data is collected. Further, the aims of the analysis will dictate which network statistics are used to analyse the social networks produced.

Step 2: define task(s) or scenario under analysis. The next step involves clearly defining the task or scenario under analysis. It is recommended that the task be described clearly, including the different agents (human and non-human) involved, the task goals, and the environment within which the task is to take place. Hierarchical Task Analysis (HTA) may be useful for this purpose if there is sufficient time available. It is important to clearly define the task as it may be useful to produce social networks for different task phases, such as pre-, during, and post accident.

Step 3: collect data. The next step involves collecting the data that is to be used to construct the social networks. This typically involves observing or recording the task under analysis and recording the links of interest that occur during task performance; however, when the focus is on accident scenarios it is highly likely that the data will be obtained from incident or inquiry reports, questionnaires and/or interviews with those involved. In this case the analyst should use the data available to identify all communications between all agents involved. Typically, the direction (i.e. from agent A to agent B), frequency, type, and content of communications are recorded. It is important to note that communications

CHAPTER 2 • HUMAN FACTORS METHODS FOR ACCIDENT ANALYSIS   49

between both human and non-human agents are often important in modern day sociotechnical systems. For example, an audible alarm communicating a warning to a control room operator or pilot, or a reference to a procedure detailing a response to a particular situation could both be classed as communications during accident analysis efforts.

Step 4: review and validate data collected. Often it is useful to revisit the data to check for errors or missing data, particularly when the data are derived from incident reports. Often all communications/associations are not recorded initially and so this step is critical in ensuring that the data are accurate. It may also be pertinent for reliability purposes to get another analyst to analyse the data in order to compute reliability statistics. Finally, it is useful to get an appropriate SME to review the data.

Step 5: construct agent association matrix. Once the data is checked and validated, the data analysis phase can begin. The first step involves the construction of an agent association matrix which depicts the frequency and direction of communications links between all agents involved in the scenario under analysis. This involves constructing a simple matrix and entering the frequency of communications between each of the agents involved. For example purposes, a simple association matrix, depicting the communications between selected cockpit and air traffic control tower agents during the events leading up to the 1977 Tenerife air disaster, is presented in Table 2-6.

Table 2-6   Social network agent association matrix example

Agent	A	B	C	D	E	F
PANAM Captain (A)	-	20	10	-	-	-
PANAM Co-Pilot (B)	18	-	2	-	-	-
PANAM Flight Engineer (C)	8	1	-	-	-	-
PANAM Radio (D)	-	-	-	-	-	9
KLM Radio (E)	-	-	-	-	-	5
Tower (F)	-	-	-	7	6	-

Step 6: construct social network diagram. Next, a social network diagram should be constructed. The social network diagram depicts each agent (both human and non-human) in the network and the associations that occurred between them during the scenario under analysis. Within the social network diagram, associations between agents are represented by directional arrows linking the agents involved. The frequency of associations is represented numerically and via the thickness of the arrows. For example purposes, a social network diagram for the Tenerife air disaster communications described in Table 2-6 is presented in Figure 2-11.

Step 7: analyse network mathematically. If appropriate, the network can be analysed using social network analysis metrics. Various metrics exist for this purpose, with those used being dependent on the analysis aims. In the past, we have found sociometric status, centrality, and density useful. For example, sociometric status provides a measure of how 'busy' a node is relative to the total number of nodes present within the network under analysis (Houghton et al. 2006). Thus, sociometric status gives an indication of the relative prominence of agents based on their links to other agents in the network. In accident scenarios, high sociometric status values may be indicative of agents who were overloaded with communications and represented a bottleneck. Centrality is also a metric of the standing of a node within a network (Houghton et al. 2006), but here this standing is in terms of its 'distance' from all other nodes in the network. A central node is one that is close to all other nodes in the network and a message conveyed from that node to an arbitrarily selected other node in the network would, on average, arrive via the least number of relaying hops (Houghton et al. 2006). In accident scenarios, agents with low centrality scores would generally be on the periphery in terms of communications, and may not have been exposed to important information or communications. Finally, network density provides an indication of how dense, in terms of associations between agents, a particular network is. Various software tools are available for analysing social networks, such as Agna and WESTT (Houghton et al. 2008).

Step 8: use network to make judgements on communications during accident scenario. The final step involves using the SNA outputs (e.g. SNA diagram, network analysis) to make judgements on the role of communications in the accident under analysis. It is useful to use SMEs during this step. Elements to look for include communications breakdowns (i.e. communications links that should have been present but were not), overloaded agents and bottlenecks, inappropriate communications (i.e. communications links that should not have been present) and also failures of communications technologies. This form of analysis is particularly powerful when an accompanying approach focusing on the content of the communications is also used.

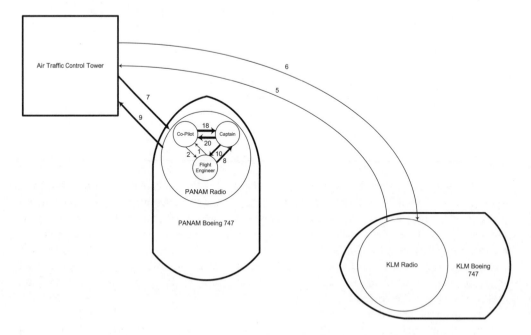

Figure 2-11   Example social network diagram for Tenerife air disaster communications

Advantages:

1. Since communications failures are typically involved in accident scenarios, SNA is particularly useful for accident analysis purposes;
2. Social network diagrams provide a powerful, and easily interpretable, means of representing the communications links between agents in complex sociotechnical systems
3. Focuses on both human and non-human agents;
4. Networks can be classified according to their structure. This is particularly useful when analysing networks across different scenarios or domains;
5. SNA is a generic approach and has been used in a range of domains for a number of different purposes;
6. Simple to learn and apply; and
7. Various free and commercial software programs are available for analysing social networks (e.g. Agna NodeXL, Pajek, NetDraw; (UCINET).

Disadvantages:

1. The data collection procedure can be time consuming;
2. When using SNA for accident analysis purposes, it is often difficult to get precise data regarding communications between agents;
3. For large, complex networks the data analysis procedure can be highly time consuming;
4. For complex collaborative tasks in which a large number of associations occur, SNA outputs can become complex and unwieldy;
5. Some knowledge of network statistics is required to interpret analysis outputs;
6. Without software support, analysing the networks mathematically is difficult, time consuming, and laborious; and
7. Focuses only on communications.

Related methods SNA typically uses observational study as its primary means of data collection, however for accident analysis purposes interviews, questionnaires or documentation review (i.e. accident inquiry report) are more likely.

Approximate training and application times SNA requires only minimal training, although some knowledge of network statistics is required to interpret the results correctly. Application time is dependent upon the task and network involved. For short tasks involving small networks with only minimal associations between agents, the application time is low, particularly if a software package is used for data analysis. For tasks of a long duration involving large, complex networks, the application time is likely to be high, due to lengthy data collection, data entry and analysis processes. Application time is reduced significantly via the use of software support, which automates the social network diagram construction and data analysis processes.

Reliability and validity Although reliability and validity is typically high, since the analysis is most often based on observable or recorded communications, when used for accident analysis purposes reliability and validity may suffer due to the input data used. In the absence of precise communications transcripts (e.g. derived from black box devices) analysts often have to estimate communications data from incident descriptions or inquiry reports.

Tools needed SNA can be conducted using pen and paper only; however, it is recommended that software support, such as the Agna, NodeXL, Pajek, or UCINET tools, be used for the network analysis component.

Example The authors recently used SNA as part of the Event Analysis of Systemic Teamwork framework (EAST; Stanton et al. 2005) to analyse the response to the World Trade Center (WTC) terrorist attacks. The emergency response, involving the New York Police Department (NYPD), the Fire Department (FDNY), the Port Authority Police Department (PAPD) and the Emergency Medical Services (EMS) was heroic and undoubtedly led to countless lives being saved. In performing the analysis we did not wish to detract from the courageous actions of those involved, rather, we felt that important insights could be derived from analysing such a complex, large-scale multi-agency response for which there was a rich source of data.

The terrorist attacks of 11 September 2001 were unprecedented in scale and nature and represented one of the greatest crises in American history (Burke 2004). At approximately 8:46am on the morning of 11 September, American Airlines Flight 11, which had been hijacked some time earlier, crashed into the North Tower of the WTC complex in New York City. Seven minutes later, at approximately 9:03am, United Airlines Flight 175, which had also been hijacked, flew directly into the WTC's South Tower. According to the National Institute of Standards and Technology, between 16,400 and 18,800 civilians were located within the WTC complex at 8.46am on 11 September (9/11 Commission 2004). The largest rescue operation in US history (McKinsey and Company 2002b) was initiated immediately after the first impact and involved primarily search and rescue operations designed to remove civilians from the North and South Towers of the WTC.

An EAST analysis of the WTC attack response involving the four main emergency organisations (NYPD, FDNY, PAPD and EMS) was undertaken based on the various accounts of the response presented in the 9/11 Commission Report (US National Commission on Terrorist Attacks upon the United States 2004), the McKinsey NYPD and FDNY/EMS response reports (McKinsey and Company 2002a, 2002b) and other academic literature. As part of the analysis SNA was used to model communications failures during the response, and the propositional network method was used to model the content of the failed communications in terms of the situation awareness being passed within and between agencies.

The procedure involved reviewing the documentation described above. Due to the size of the incident and breadth of the activities undertaken, the analysis focused on vignettes as opposed to the entire scenario; this approach is advocated by the NATO code of best practice (NATO 2002) for studying command and control systems, suggesting that an appropriate way of dealing with the complexity of such systems is for analysts to present their findings in the form of vignettes. The results from the analysis of two vignettes (potential rooftop rescue and North Tower evacuation) are presented below.

Potential rooftop rescue vignette Approximately six minutes after the first impact, an NYPD helicopter arrived at the scene and assessed the feasibility of undertaking a rooftop rescue on the North Tower. Social and propositional networks were constructed for the rooftop rescue vignette, which involved the helicopter's assessment of the potential for a rooftop rescue, and the communication of this to NYPD chiefs. The social and propositional networks are presented in Figure 2-12. Within the figure the arrows linking agents denote the communications between them, and the accompanying propositional networks represent the content of the situation awareness exchanges (in terms of information elements and the relationships between them) occurring as a result of the communications.

As the vignette presented in Figure 2-12 illustrates, the results of the initial rooftop assessments were not communicated across agencies; rather they were passed onto the NYPD chief who then sent out an order to NYPD officers that a rooftop rescue should not be attempted. This information was not passed onto the FDNY (who by protocol should have had a representative in any aviation unit assessing the feasibility of a rooftop rescue) nor was it passed onto 911 operators and FDNY dispatchers. As a corollary, trapped civilians (some advancing toward the rooftop) in the North Tower who were calling 911 for advice on whether to ascend or descend could not be advised of the lack of a rooftop rescue, since 911 operators and FDNY dispatchers were not aware of the 'no rooftop rescue' order.

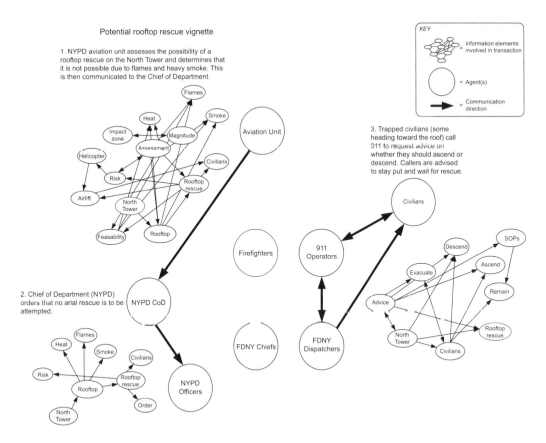

Figure 2-12 Potential rooftop rescue vignette

North Tower evacuation vignette At around 9:59am the South Tower collapsed. Civilians and rescue personnel located within the North Tower were not initially aware of this. The following vignette relates to the evacuation, by emergency service personnel, from the North Tower in response to the collapse of the South Tower. Social and knowledge networks were developed for each agency involved. The social and knowledge networks for the NYPD and FDNY's evacuation from the North Tower are presented in Figure 2-13 and Figure 2-14.

As both figures demonstrate, the NYPD and FDNY social and knowledge networks for the North Tower evacuations were very different. The NYPD were immediately informed by their aviation unit that the South Tower had collapsed, that the top 15 floors of the North Tower were glowing red and that the pilot felt that the North Tower would not stand for much longer. In response to this the ESU (Emergency Service Unit) officer at the command post reported to all ESU units in the North Tower that the South Tower had collapsed and issued an evacuation order. The evacuating units then passed this information onto any fire fighters and civilians that they came across when evacuating. In contrast, the FDNYs social and propositional networks (Figure 2-14) show that the FDNY personnel within the North Tower were not aware that the South Tower had collapsed; rather, upon hearing a violent roar and seeing debris, smoke and the emergency lights

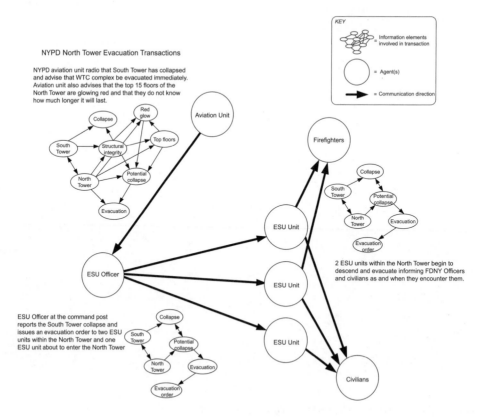

Figure 2-13 NYPD North Tower evacuation vignette; figure shows the social network involved and the situation awareness exchanges between parties

CHAPTER 2 • HUMAN FACTORS METHODS FOR ACCIDENT ANALYSIS   55

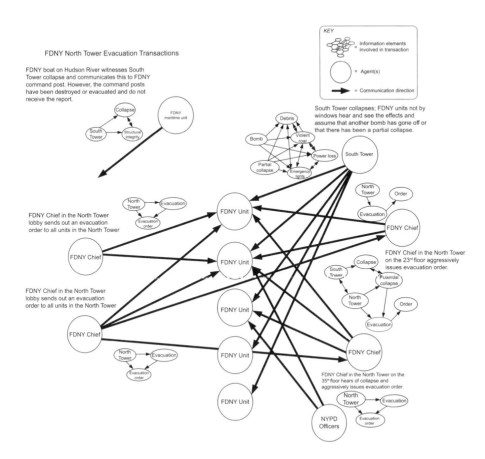

Figure 2-14   FDNY North Tower evacuation vignette; figure shows the social network involved and the situation awareness exchanges between parties

being activated, most assumed that a bomb had gone off or that there had been a partial collapse (9/11 Commission 2004). A FDNY boat on the River Hudson had witnessed the South Tower collapse and had attempted to pass on this information to the FNDY command post, but this had been destroyed and evacuated in the collapse and so this information was not communicated. Despite this, two FDNY chiefs within the North Tower issued an evacuation order, but crucially these orders did not contain the fact that the South Tower had collapsed (9/11 Commission 2004) and also some fire fighters did not receive the orders due to problems with the radios. Finally, a FDNY chief on the 35th floor heard that the South Tower had collapsed and aggressively initiated an evacuation.

Perhaps unsurprisingly due to the novelty, complexity and scale of the emergency, the exchange of information between the different agencies involved in the WTC response was often not efficient; as the 9/11 Commission report concludes, 'the incident command system did not function to integrate awareness among agencies or to facilitate interagency response'. As a result the level of coordination between agencies during the response was not optimal (9/11 Commission 2004). The analysis presented identified instances where communications between the responding agencies was sub-optimal, which led to poor exchanges of situation awareness between agencies. These included instances where a communication was required but was not initiated, instances where a communication between agencies was initiated as required but was not completed for some reason or other, erroneous communications, where the content of the communication between parties was incorrect, inappropriate communications, whereby inappropriate or wrong information was exchanged between parties, and finally conflicting communications, whereby one agency was receiving conflicting information from other parts of the system. Examples of each communications failure type are presented in Table 2-7.

Table 2-7   Communication failure types

Transaction Failure Type	Example
Communication not initiated	• The different agencies involved did not pass on which floors had been searched for civilians; • FDNY chiefs ordered both towers to be evacuated at 8:57am; this was not conveyed to 911 operators and FDNY dispatchers who subsequently advised civilians in the Towers to stay put and wait for rescue; • 911 operators and FDNY dispatchers were not informed that the possibility of a rooftop rescue on the North Tower had been discounted by the NYPD aviation unit; • NYPD helicopter did not communicate South Tower collapse or the imminent North Tower collapse to any other agencies.
Communication initiated but not completed	• FDNY boat on the Hudson River communicated collapse of the South Tower on tactical channel but no FDNY personnel received the communication; • A group of civilians trapped on the 105th floor of the North Tower called 911 to inform them of their location; this information was passed onto FDNY dispatchers who passed it onto a field comms operator who then passed it onto the outdoor command post, who attempted to pass it onto the chiefs in the North Tower lobby but failed to do so; • FDNY field communications van attempted to track responding units on a magnetic whiteboard but were overwhelmed by the number of units; • One group of callers trapped on the 83rd floor repeatedly called 911 to clarify whether the fire was above or below them; they were transferred between operators and put on hold but were never answered (evidence suggests that they died).
Erroneous communications	• When the South Tower collapsed, firefighters and ESU units located in the North Tower heard a violent roar, were knocked off their feet, saw debris and assumed that a bomb had gone off or that a partial collapse had occurred.
Inappropriate communications	• FDNY chiefs informed FDNY dispatch operators that the North and South Towers were not at risk of a total collapse; • The South Tower public address system announced that the building was safe and that workers should remain in their offices; subsequently workers in the process of evacuating stopped and went back to work.
Conflicting communications	• Various civilians were told to evacuate but also to stay put and await rescue; e.g. a group of Port Authority employees on the 64th floor were told (via a third party in telephone contact with them) by the Port Authority Police (Newark Airport desk) to evacuate; however, upon direct contact with the Port Authority Police (Jersey City desk) they were told to stay put and wait for the police to arrive. The third party contacted the Port Authority (Newark Airport desk) once more who advised that the group should evacuate. The workers were in fact not trapped and, despite eventually evacuating, most died in the North Tower collapse.

2.3.7 Propositional Networks

Background and applications The propositional network methodology was originally developed for modelling Distributed Situation Awareness (DSA; Salmon et al. 2009) in complex sociotechnical systems. Based on the notion that knowledge comprises concepts and the relationships between them (Shadbolt and Burton 1995), propositional networks use networks of linked information elements to depict a system's awareness, including the information underpinning DSA and the relationships between the different pieces of information. Importantly propositional networks consider both human and non-human agents (e.g. displays, technologies, procedures), purporting that a systems awareness is distributed across the different agents, both human and non-human (i.e. displays, manuals, tools), comprising the system. Once completed, the structure and content of the networks are typically analysed using network analysis metrics. For accident analysis purposes, propositional networks are useful for identifying situation awareness failures that played a role in accidents.

Domain of application Propositional networks were originally applied for modelling DSA in command and control scenarios in the military (e.g. Stanton et al. 2006; Salmon et al. 2009) and civilian (e.g. Salmon et al. 2008) domains; however, the method is generic and can be applied in any domain. The method has since been applied in a range of domains, including naval warfare (Stanton et al. 2006), land warfare (Salmon et al. 2009) railway maintenance operations (Walker et al. 2006), road transport (Walker et al. 2009), and military aviation airborne early warning systems (Stewart et al. 2008).

Accident analysis/investigation applications Propositional networks have recently been applied for accident analysis purposes in the context of civil and military aviation accidents (e.g. Griffin et al. 2010; Rafferty et al. forthcoming). For example, Griffin et al. (2010) used propositional networks to model the Kegworth British Midland Flight 92 air disaster in which a Boeing 737-400 crash landed on an embankment of a motorway near Kegworth in the UK following an inappropriate engine shut down.

Procedure and advice Step 1: define analysis aims. Firstly, the aims of the analysis should be clearly defined, since this affects the scenarios used and the propositional networks developed. When used for accident analysis purposes, a specific scenario has normally been identified and the aims of the analysis typically centre on the situation awareness of the system and its components (human and non-human agents) and communication of information prior to and during the accident in question.

Step 2: collect data regarding the accident scenario under analysis. Propositional networks can be constructed from a variety of data sources. These include observational study and/or verbal transcript data, CDM data, HTA data or data derived from work-related artefacts such as Standard Operating Instructions (SOIs) or procedures, user manuals, and training manuals. When using propositional networks for accident analysis purposes, detailed data regarding the accident under analysis should be collected. This might involve reviewing official accident or inquiry reports, conducting interviews with personnel involved in the accident, or gathering SME accounts of the accident.

Step 3: define task phases. It is normally useful to identify distinct task or scenario phases. This is often done based on distinct task or time-based phases. This allows propositional networks to be developed for each phase, which is useful for depicting the dynamic and changing nature of DSA throughout a task or scenario.

Step 4: define concepts and relationships between them. In order to construct propositional networks, firstly the concepts need to be defined followed by the relationships between them. For the purposes of DSA assessments, the term 'information elements' is used to refer to concepts. To identify the information elements related to the task under analysis, a simple content analysis is performed on the input data and keywords are extracted. These keywords represent the information elements, which are then linked based on their causal links during the activities in question (e.g. cockpit 'has' display, pilot 'knows' airspeed). Links are represented by directional arrows and should be overlaid with the linking proposition. A simplistic example of the relationship between concepts is presented in Figure 2-15.

The output of this process is a network of linked information elements; the network contains all of the information that is used by the different agents during task performance and depicts the relationships between the information elements. The network thus represents the system's awareness, or what the system 'needed to know' in order to successfully undertake task performance.

Step 5: define information element usage/contribution. It is normally useful to define and represent different agents' usage of the information elements contained in the networks. This is represented via shading of the different nodes within the network based on their usage/contribution by different agents during task performance. During this step the analyst identifies which information elements the different agents involved used or contributed during task performance. This can be done in a variety of ways, including by further analysing input data, and/or by holding discussions with those involved or relevant SMEs.

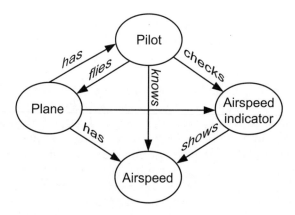

Figure 2-15   Example relationships between concepts

Step 6: identify network failures/breakdowns. When using propositional networks for accident analysis purposes, one typically focuses on situation awareness failures. At this stage it is useful for the analyst to use the input data to identify the following situation awareness failures:

1. Loss of situation awareness;
2. Situation awareness incomplete/inaccurate (i.e. missing or erroneous information elements);
3. Situation awareness-related information not communicated;
4. Erroneous understanding of information (i.e. misunderstanding of information elements);
5. Failure to integrate information appropriately; and
6. Incompatible portions of situation awareness across agents/teams/agencies.

Step 7: review and refine network. Constructing propositional networks is a highly iterative process that normally requires numerous reviews and re-iterations. It is recommended that once a draft network is created, it is subject to at least three reviews. It is useful to involve domain SMEs, or the participants who performed the task, in this process. The review normally involves checking the information elements and the links between them and also the classification of information element usage. Re-iterations to the networks normally include the addition of new information elements and links, revision of existing information elements and links and also modifying the information element usage based on SME opinion.

Step 8: analyse networks mathematically. Depending on the analysis aims and requirements, it may also be pertinent to analyse the propositional networks mathematically using network statistics. For example, it may be that a piece of information critical to the system's DSA was not communicated by a particular display or agent, or was misunderstood by a particular agent during the accident scenario. In the past we have used sociometric status and centrality calculations to identify the 'key' information elements within propositional networks. Sociometric status provides a measure of how 'busy' a node is relative to the total number of nodes present within the network under analysis (Houghton et al. 2006). In this case, sociometric status gives an indication of the relative prominence of information elements based on their links to other information elements in the network. Centrality is also a metric of the standing of a node within a network (Houghton et al. 2006), but here this standing is in terms of its 'distance' from all other nodes in the network. A central node is one that is close to all other nodes in the network and a message conveyed from that node to an arbitrarily selected other node in the network would, on average, arrive via the least number of relaying hops (Houghton et al. 2006). Key information elements are defined as those that have salience for each scenario phase, salience being defined as those information elements that act as hubs to other information elements. Typically, those information elements with a sociometric status value above the mean sociometric status value, and a centrality score above the mean centrality value, are identified as key information elements. Other network metrics, such as density and diameter have also been used to interrogate situation awareness networks (e.g. Walker et al. 2011).

Advantages:

1. Propositional networks depict the information elements underlying a system's DSA and the relationships between them;
2. Breakdowns in situation awareness can be identified and represented in the network, allowing identification of the role of situation awareness failures in accidents;
3. In addition to modelling the systems awareness, propositional networks also depict the awareness of individuals and sub-teams working within the system;
4. The networks can be analysed mathematically in order to identify the key pieces of information underlying a system's awareness. This is useful for identifying when personnel were not aware of key pieces of information;
5. Unlike other situation awareness measurement methods, propositional networks consider the mapping between the information elements underlying situation awareness;
6. The propositional network procedure avoids some of the flaws typically associated with situation awareness measurement methods, including intrusiveness, high levels of preparatory work (e.g. situation awareness requirements analysis, development of probes) and the problems associated with collecting subjective situation awareness data post trial;
7. Propositional networks are easy to learn and use; and
8. Software support is available via the Leximancer and WESTT software tools (see Houghton et al. 2008 and Walker et al. 2011).

Disadvantages:

1. Constructing propositional networks for complex accident scenarios can be highly time consuming and laborious;
2. Analysing the networks mathematically adds further time to the analysis and can also be laborious;
3. It is difficult to present larger networks within articles, reports, and/or presentations;
4. The initial data collection phase may involve a series of activities and often adds considerable time to the analysis;
5. The reliability of the method is questionable, particularly when being used by inexperienced analysts without software support; and
6. For retrospective events, it is often difficult to get data that is sufficiently rich to support the construction of accurate propositional networks.

Related methods Propositional networks are similar to other network-based knowledge representation methods such as semantic networks (Eysenck and Keane 1990) and concept maps (Crandall et al, 2006). The data collection phase typically utilises a range of other Human Factors methods, including observational study, CDM interviews, verbal protocol analysis and HTA. The networks can also be analysed using various metrics derived from social network analysis methods.

Approximate training and application times Providing the analysts involved have an understanding of DSA theory, the training time required for the propositional network method is low; our experiences suggest that a day's training is sufficient. Following training, however, considerable practice is required before analysts become proficient in the method, particularly with regard to the linking of different information elements. The application time is typically high, although it can be low provided the task is simplistic and short. Application time can be reduced significantly through the use of software support such as the Leximancer textual analysis software.

Reliability and validity The content analysis procedure should ease some reliability concerns; however, the links between concepts are often made on the basis of the analyst's subjective judgement and so the reliability of the method may be limited, particularly when being used by inexperienced analysts. The validity of the method is difficult to assess, although our experiences suggest that validity is high, particularly when appropriate SMEs are involved in the process. Reliability can be enhanced through the use of software approaches, such as Leximancer, which automate the content analysis and information element linkage components.

Tools needed On a simple level, propositional networks can be conducted using pen and paper; however, a drawing package such as Microsoft Visio is often used to construct the propositional networks. Various forms of software support are also available. For example, the Leximancer software tool automates the content analysis and information element linkage processes. Houghton et al. (2008) also describe the WESTT software tool, which contains a propositional network construction module that auto-constructs propositional networks based on text data entry. Finally, the Agna network analysis software tool is useful for analysing the networks mathematically.

Example output To demonstrate how the propositional network approach can be used for accident analysis efforts, a simple analysis of the *Herald of Free Enterprise* ferry disaster is presented. The *Herald of Free Enterprise* passenger ferry capsized in shallow waters just outside Zeebrugge harbour on 6 March 1987, killing 150 passengers and 38 crew members. The immediate cause of the disaster, the flooding of the lower deck, was attributed to the ferry setting sail with its inner bow doors open, which was caused by the assistant bosun's failure to close the ship's bow doors and the captain's decision to set sail with the bow doors open (although the captain was not aware of this). Various contributory factors have since been identified (see Reason 1990 for a full description), including the assistant bosun's level of fatigue, poor rostering (the assistant bosun was in fact asleep in his cabin, having recently been relieved from other additional duties), the bosun's failure to shut the bow doors even though he actually recognised that they were open (he felt that it was not his job to shut them), pressure on the crew to depart early due to delays at Dover and the 'choppy' sea conditions on the day of the disaster. There were also a number of board and policy failures, including a negative reporting culture within the company which ensured that the master did not know of any problems encountered previously, the failure to install a bow door indicator on the bridge despite repeated requests, and the unsafe top-heavy design of the ferry involved (Reason 1990).

The propositional network approach can be used in this case to highlight how situation awareness failures contributed to the accident. Figure 2-16 presents four extracts of networks for the scenario, the first representing the actual scenario, and the remaining three representing scenarios in which the accident would have been prevented given the state of the systems awareness that is presented. In the network extracts, the information elements are shaded in order to represent who in the system is aware of them (an element with no shading represents information that none of the personnel involved were aware of). The actual scenario network extract (top of Figure 2-16) shows how, at the point of setting sail, the ship's captain and the bosun were under the impression that the bow doors were closed (the bosun had seen them open but assumed that the assistant bosun would close them). Three scenarios, in which the systems awareness was different and the accident could have prevented follow. In prevention scenario 1, the bosun sees the bow doors open, and closes them accordingly. Unfortunately, a culture of not-my-job meant that the bosun did not do so during the actual accident. In prevention scenario 2, the bosun is aware that the assistant bosun is asleep and closes the bow doors for him. In prevention scenario 3, the assistant bosun hears the harbour station's call and closes the bow doors accordingly. Unfortunately, at the time of departing the assistant bosun was asleep due to having been relieved from other duties. Finally, in prevention scenario 4, the ship's captain becomes aware that the bow doors are open (through an appropriate display on the bridge) and communicates with the bosun an order to close them. Unfortunately, this scenario was prevented primarily due to the refusal by company management to install a bow door indicator on the bridge, despite repeated requests from the ship's captain (Reason 1990). The propositional network approach is therefore useful for highlighting how accidents could have been avoided in terms of the level of situation awareness held by the system. In this case, a not-my-job culture, poor rostering, and lack of appropriate technologies on the bridge are all factors which diminished the chances of the accident being avoided.

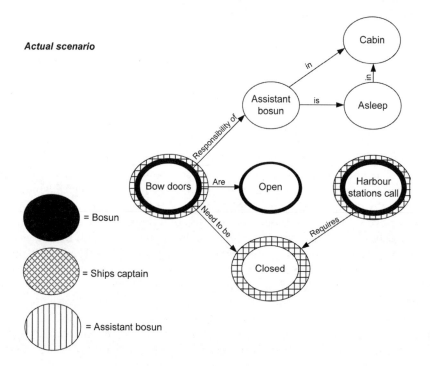

Figure 2-16 *Herald of Free Enterprise* incident; actual and prevention networks

CHAPTER 2 • HUMAN FACTORS METHODS FOR ACCIDENT ANALYSIS 63

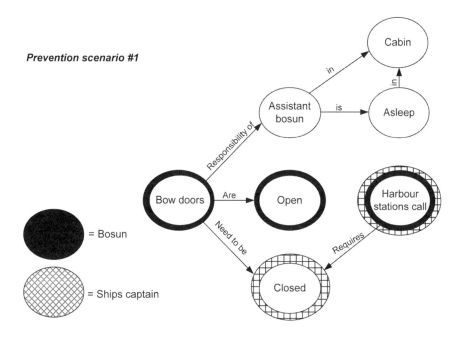

Figure 2-16 *Continued*

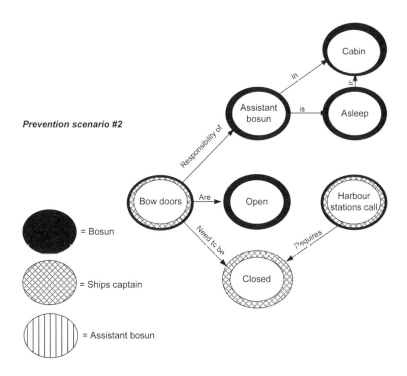

Figure 2-16 *Continued*

Prevention scenario #3

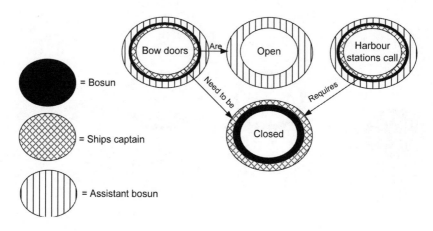

Figure 2-16 *Continued*

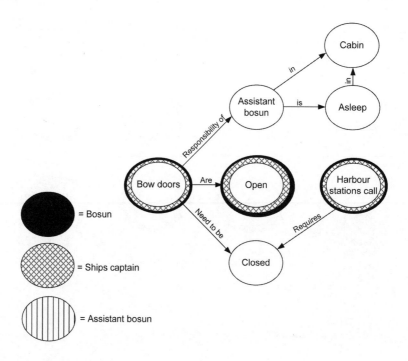

Figure 2-16 *Concluded*

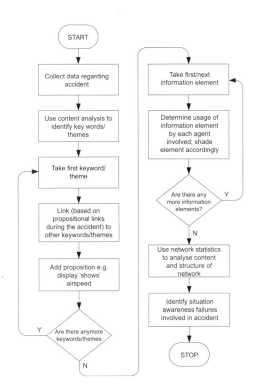

Flowchart 2-4   Propositional networks

Recommended Reading

Griffin, T.G.C., Young, M.S., Stanton, N.A. (2010) Investigating accident causation through information network modelling. *Ergonomics*, 53:2, 198–210.

Salmon, P.M., Stanton, N.A., Walker, G.H. and Jenkins, D.P. (2009) *Distributed Situation Awareness: Advances in Theory, Modelling and Application to Teamwork*. Aldershot, UK: Ashgate Publishing.

2.3.8 Critical Path Analysis

Background and applications   Critical Path Analysis (Baber 2004) is used to model the performance time associated with sequences of tasks. Accordingly, CPA has previously been used to model appropriate operator response times during accident scenarios (e.g. Stanton and Baber 2008). CPA works by modelling task sequences and calculating, using standard human task performance time data, the time required for completion of different sequences of tasks.

Domain of application   CPA was originally developed as a project management tool but has been mainly used by Human Factors practitioners in the realm of Human Computer Interaction (HCI).

Accident analysis/investigation applications Stanton and Baber (2008) recently used CPA to model the signaller's response time during the Ladbroke Grove rail crash (see Chapter 7). This was then compared to the actual response time during the accident itself to denote whether the signaller's response to the 'signal passed at danger alarm' was appropriate.

Procedure and advice Step 1: define task(s). The first step involves defining the tasks to be analysed. When used for accident analysis purposes this involves reviewing data regarding the accident and constructing a detailed HTA for the scenario under analysis. The task description should be as detailed as possible and include specific sub-tasks. For example, for the activity 'accessing an automated teller machine (ATM)', Baber (2004) uses the following task list: (1) retrieve card from wallet; (2) insert card into ATM; (3) recall PIN; (4) wait for screen to change; (5) read prompt; (6) type in PIN digit; (7) listen for confirmatory beep; (8) repeat Steps 6–7 until full PIN has been entered; (9) wait for screen to change. It is important at this stage to describe the task in sufficient detail so that task times can be assigned to component tasks.

Step 2: define task input and output modalities. Each component task step should be defined in terms of the input and output modalities involved, that is, which sensory modality is used for input (e.g. entering PIN digit) and which is used for the output (e.g. listening to confirmatory beep). The following input/output modalities are used (Baber 2004): manual (left and right hand), visual, auditory, cognitive, speech. For the ATM example described in Step 1, the input and output modalities are presented in Table 2-8.

Step 3: construct task sequence/dependency chart. The next step involves constructing a task sequence/dependency chart. This shows the possible ordering and sequence of the tasks and the links between them. An example task sequence/dependency chart for the ATM example is presented in Figure 2-17.

Table 2-8   Input/output modalities for ATM example (adapted from Baber, 2004)

Task step	Manual (L)	Manual (R)	Speech	Auditory	Visual	Cognitive	System
Retrieve card	■						
Insert card		■					
Recall PIN						■	
Screen change							■
Read prompt					■		
Type digit		■					
Listen for beep				■			
Screen change							■

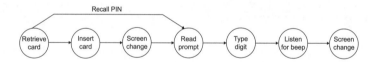

Figure 2-17   Task sequence/dependency chart for ATM example (adapted from Baber 2004)

Step 4: assign task times to each task step. The next step involves assigning appropriate task times to each task step involved. Standard task completion times are typically sourced from the HCI literature (Stanton and Baber 2008). For example, the Keystroke Level Model method (see Stanton et al. 2005 or Stanton and Young 1999) has a range of associated task completion times for HCI type tasks. The task completion times for the ATM example are presented in Table 2-9.

Step 5: calculate forward pass. Initially a forward pass is calculated. For the automated teller example, Baber (2004) suggests that the analyst should begin at the first task and assign an earliest start time of zero. The finish time will be zero plus the duration of the task step, e.g. 'retrieve card' takes 500ms, so the earliest finish time will be 500ms. The earliest finish time of one task becomes the earliest start time for the next task (see Table 2-9).

Step 6: calculate backward pass. Next, the backward pass is calculated. This involves beginning at the last node and assigning a latest finish time. To calculate the latest start time, the analyst should subtract the task duration from the latest finish time (see Table 2-9).

Step 7: calculate critical path and total task completion time. The critical path comprises all nodes that have zero difference between earliest finish time and latest finish time (Baber 2004). For the ATM example, the task step 'recall PIN' has a 320ms float, which means that it can be initiated up to 320ms after the other tasks on the critical path without having an impact on overall performance. For accident analysis purposes, it is likely that the total task completion time will be of interest. In this case, the total task completion time can be found by tracing through the CPA using the longest node-to-node values.

Step 8: compare CPA task performance time to actual task performance time. When using CPA for accident analysis purposes, the final step involves comparing the task performance time calculated through CPA to the actual task performance times incurred during the accident under analysis.

Table 2-9   Critical path calculation table

Task step	Duration	Earliest start	Latest start	Earliest finish	Latest finish	Float
Retrieve card	500ms	0	0	500	500	0
Insert card	350ms	500	500	850	850	0
Recall PIN	780ms	0	320	780	1100	320
Screen change	250ms	850	850	1100	1100	0
Read prompt	350ms	1100	1100	1450	1450	0
Type digit	180ms	1450	1450	1630	1630	0
Listen for beep	100ms	1630	1630	1730	1730	0

Advantages:

1. Can be used to model appropriate task performance times during accident scenarios;
2. Is able to model parallel tasks;
3. Simple to use requiring minimal training;
4. Standard unit task completion times are available in the literature (e.g. Stanton and Young 1999); and
5. Has been used previously for accident analysis purposes (e.g. Stanton and Baber 2008).

Disadvantages:

1. Could be used to support allocation of blame to human operators;
2. Does not identify or consider causal factors from the wider organisational system;
3. Can only model error free performance so does not consider performance shaping factors likely to increase operator task performance times;
4. Modality can be difficult to define;
5. Identifying appropriate task time data for complex tasks could be difficult;
6. Detailed task analysis is required; and
7. Can be tedious and time consuming for complex tasks.

Related methods   CPA typically uses an HTA (Stanton 2006) of the task under analysis as its input. Other task performance time modelling approaches also exist, such as the Keystroke Level Model (Card et al. 1983).

Approximate training and application times   The training time for CPA is low. Typically, application time is also low; however, this is dependent upon the data available and also the complexity of the task under analysis. For example, for tasks that are complex and have no associated HTA description, the application time is likely to be high due to the requirement to construct an initial HTA.

Reliability and validity   There is limited reliability and validity evidence presented in the literature; however, Baber and Mellor (2001) report encouraging validity for predicted task times in user trials of the method. The reliability of the method is likely to be high if an existing HTA is available; however, for complex tasks where an HTA is not available reliability may be questionable.

Tools needed   CPA can be applied using pen and paper only; however, it is normally useful to draw CPA outputs using Microsoft Visio or Word.

CHAPTER 2 • HUMAN FACTORS METHODS FOR ACCIDENT ANALYSIS

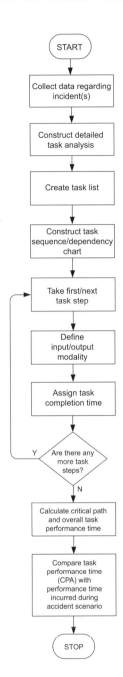

Flowchart 2-5   Critical path analaysis

Recommended Reading

Baber, C. (2004) Critical path analysis. In: N.A. Stanton, A. Hedge, K. Brookhuis, E. Salas and H. Hendrick (eds), *Handbook of Human Factors and Ergonomics Methods*. Boca Raton, FL: CRC Press.

Baber, C. and Mellor, B.A. (2001) Modelling multimodal human-computer interaction using critical path analysis. *International Journal of Human Computer Studies*, 54, 613–36.

2.3.9 The Technique for the Retrospective and Predictive Analysis of Cognitive Errors

Background and applications   The Technique for the Retrospective and Predictive Analysis of Cognitive Errors (TRACEr; Shorrock and Kirwan 2002) was developed for predicting or retrospectively analysing air traffic controller errors. The method has two components: an error prediction component and a retrospective incident analysis component (only the retrospective analysis component is discussed here; for a description of the predictive component readers are referred to Shorrock and Kirwan (2002)). TRACEr uses the following six error taxonomies that were developed based on a series of activities, including a literature review, an analysis of air traffic control incidents, and interviews with air traffic controllers (Shorrock and Kirwan 2002): Task Error, Information; Performance Shaping Factors (PSFs); External Error Modes (EEMs); Internal Error Modes (IEMs); and Psychological Error Mechanisms (PEMs). In addition, the classification of error detection and recovery strategies is supported by a series of pre-defined questions.

Domain of application   TRACEr was originally developed for use in air traffic control; however, the majority of its taxonomies are generic, allowing it to be applied in other safety-critical domains.

Accident analysis/investigation applications   TRACEr has mainly been applied for accident/incident analysis purposes within the air traffic control domain (e.g. Shorrock and Kirwan 2002); however, recently the method has also been applied to the analysis of rail incidents (e.g. Baysari et al. 2009).

Procedure and advice (retrospective analysis only)   Step 1: data collection. TRACEr is dependent upon accurate data regarding the incident(s) under analysis. The first step therefore involves collecting detailed data regarding the cases to be analysed. Although it can be used for single cases, in the past TRACEr has predominantly been used to analyse multiple incident cases (e.g. Shorrock and Kirwan 2002; Baysari et al. 2009), so this might involve gathering data from incident/accident databases or combining data sets. Accident databases typically include a narrative of each incident along with some judgement on the causal factors involved; however, other useful data might also be available, including video recordings of the incident, investigation reports, and accounts from those involved.

Step 2: identify error events. TRACEr analyses are based on the initial identification of 'error events'. Each error event is then analysed using the TRACEr taxonomies. Initially the error events involved are identified. This involves reviewing the data surrounding the

CHAPTER 2 • HUMAN FACTORS METHODS FOR ACCIDENT ANALYSIS

incident and identifying and recording each error event. Error event identification can be undertaken using analyst subjective judgement; however, it is useful to involve SMEs or even the operators involved in the incident under analysis.

Step 3: identify task error involved. For each error event, the task error involved is first identified. The task error component describes the task that was being performed when the error of interest was made, e.g. radar monitoring error. Examples of the TRACEr task errors are presented in Table 2-10 and include 'radar monitoring error', 'co-ordination error' and 'flight progress strip use error' (Shorrock and Kirwan 2002).

Step 4: identify internal error mode involved. Next, the analyst has to identify the Internal Error Mode (IEM) associated with the error. The IEM describes which cognitive function failed (Shorrock and Kirwan 2002), with examples including late detection, misidentification, hearback error, forget previous actions, prospective memory failure, mis-recall stored information and mis-projection. During this step the information taxonomy is also used to describe the 'subject matter' of the error in terms of the information that was involved. For example, certain information might have been misunderstood, misjudged or subject to perceptual failure. Categories of information considered by TRACEr include controller materials (e.g. flight progress strips), controller activities (e.g. transfer), variable aircraft information (e.g. aircraft call sign), time and location (e.g. airspace type), and airport (e.g. runway) (Shorrock and Kirwan 2002).

Step 5: identify psychological error mechanism(s) involved. The Psychological Error Mechanism (PEM) component is used to identify the psychological failures that led to the error. Here the analyst uses the TRACEr PEM taxonomy to identify the PEM involved. The PEM taxonomy is presented in Table 2-12.

Table 2-10   Extracts from TRACEr's task error and performance shaping factors taxonomy (adapted from Shorrock and Kirwa, 2002)

Task error examples	
Separation error	Controller-pilot communication error
Radar monitoring error	Aircraft observation/recognition error
Co-ordination error	Control room communication error
Aircraft transfer error	Hand over/take-over error
Flight progress strip use error	Operational materials checking error
Training, supervision or examining error	Human machine interaction error
Performance shaping factors taxonomy examples	
Performance shaping factor category	**Example performance shaping factor**
Traffic and airspace	Traffic complexity
Pilot/controller communications	Workload
Procedures	Accuracy
Training and experience	Task familiarity
Workplace design, human machine interaction and equipment factors	Radar display
Ambient environment	Noise
Personal factors	Alertness/fatigue
Social and team factors	Handover/takeover
Organisational factors	Conditions of work

Table 2-11 TRACEr external error mode taxonomy (adapted from Shorrock and Kirwan 2002)

External error modes		
Selection and quality	Timing and sequence	Communication
Omission	Action too long	Unclear information transmitted
Action too much/too little	Action too short	Unclear information recorded
Action in wrong direction	Action too early	Information not sought/obtained
Wrong action on right object	Action too late	Information not transmitted
Right action on wrong object	Action repeated	Information not recorded
Wrong action on wrong object	Misordering	Incomplete information transmitted
Extraneous act		Incomplete information recorded
		Incorrect information transmitted
		Incorrect information recorded

Step 6: identify performance shaping factors involved. The next step involves identifying any Performance Shaping Factors (PSFs) involved in the incident. These represent factors that influenced human operator performance in a way that caused the error of interest to be made, and include nine categories of PSF. Each PSF category and example PSFs are presented in Table 2-10.

Step 7: identification of error detection and error correction. The final step involves identifying how the error was detected and also any error correction or recovery strategies that were employed by the controller. Here the analyst uses a series of pre-defined prompts to determine how the error was identified by the controller and also how the controller responded to the error in terms of correction or recovery strategies employed. The error detection component is supported by the following prompts (Shorrock and Kirwan 2002):

1. How did the controller become aware of the error? (e.g. action feedback, inner feedback, outcome feedback);
2. What was the feedback medium? (e.g. radio, radar display);
3. Did any factors, internal or external to the controller, improve or degrade the detection of the error?; and
4. What was the separation status at the time of error detection?

Next, the analyst uses the following prompts to determine the error correction strategy employed by the controller(s) involved:
1. What did the controller do to correct the error? (e.g. reversal or direct correction, automated correction);
2. How did the controller correct the error? (e.g. turn or climb);
3. Did any factors, internal or external to the controller, improve or degrade the detection of the error?; and
4. What was the separation status at the time of the error correction?

Steps 2–6 should be repeated for each of the error events involved in the accident.

Step 8: calculate inter-rater reliability statistics and resolve disagreements. When more than one analyst is used, it is accepted practice to calculate inter-rater reliability statistics. This normally involves the use of standard reliability tests such as Cohen's Kappa or signal detection theory sensitivity index calculations. Also at this stage any disagreements between analysts are resolved through further discussion until consensus is reached.

Step 9: analyse outputs using frequency counts. When multiple accident cases are analysed, simple frequency counts are used initially to derive an overview of the analysis. This involves calculating the frequency with which the different errors and associated factors (e.g. PEMs, IEMS, PSFs) are involved in the accidents analysed.

Table 2-12   Internal error modes and psychological error mechanisms (adapted from Shorrock and Kirwan 2002)

Internal error modes	Psychological error mechanisms
Perception	
No detection (visual)	Expectation bias
Late detection (visual)	Spatial confusion
Misread	Perceptual confusion
Visual misrepresentation	Perceptual discrimination failure
Misidentification	Perceptual tunneling
No identification	Stimulus overload
Late identification	Vigilance failure
No detection (auditory)	Distraction/preoccupation
Hearback error	
Mishear	
Late auditory recognition	
Memory	
Forget to monitor	Similarity interference
Prospective memory failure	Memory capacity overload
Forget previous actions	Negative transfer
Forget temporary information	Mislearning
Mis-recall temporary information	Insufficient learning
Forget stored information	Infrequency bias
Mis-recall stored information	Memory block
	Distraction/preoccupation
Judgment, planning and decision-making	
Misprojection	Incorrect knowledge
Poor decision	Lack of knowledge
Late decision	Failure to consider side or long-term effects
No decision	Integration failure
Poor plan	Misunderstanding

Table 2-12   *Concluded*

No plan	Cognitive function
Under-plan	False assumption
	Prioritisation failure
	Risk negation/tolerance
	Risk recognition failure
	Decision freeze
Action execution	
Selection error	Manual variability
Positioning error	Spatial confusion
Timing error	Habit intrusion
Unclear information transmitted	Perceptual confusion
Unclear information recorded	Functional confusion
Incorrect information transmitted	Dysfluency
Incorrect information recorded	Mis-articulation
Information not transmitted	Inappropriate intonation
Information not recorded	Thoughts leading to actions
	Environmental intrusion
	Other slip
	Distraction/preoccupation

Advantages:

1. Provides a comprehensive analysis of the errors involved in accident scenarios, considering the error and various associated factors (e.g. IEMs, PEMs, PSFs);
2. Has encouraging comprehensiveness and usability data (Shorrock and Kirwan 2002);
3. A less resource intensive version, TRACEr-lite, has been developed (see Shorrock 2006);
4. Can also be used predictively;
5. Has been applied in other domains (e.g. Baysari et al. 2009)
6. Although focuses heavily on individual operator errors the method also considers system-wide performance shaping factors; and
7. Was developed based on a range of activities, including analysis of accident and incident reports, interviews with SMEs and a review of the human error identification and analysis literature

Disadvantages:

1. As a by-product of its comprehensiveness, the method may seem over-complicated to some;
2. Again due to its exhaustive nature, analyses may be time-consuming;

3. Without access to the personnel involved, it will often be difficult to find the data required to support TRACEr analyses;
4. A lack of validation evidence for the method is presented in the academic literature;
5. Existing error classification approaches may be simpler and quicker to use (e.g. SHERPA);
6. Focuses almost exclusively on errors, failing to consider in detail failures across the wider organisational system; and
7. Since TRACEr was developed specifically for air traffic control some of its taxonomies/prompts are inapplicable to other domains.

Related methods TRACEr is a taxonomy-based error prediction and analysis method, many of which exist. For example, the Cognitive Reliability and Error Analysis Method (CREAM; Hollnagel 1998) and the Systematic Human Error Prediction and Reduction Approach (SHERPA; Embrey 1986) are two other popular taxonomy-based approaches. Task analysis methods, such as Hierarchical Task Analysis (Stanton 2006) are also typically used for the task description component of TRACEr.

Approximate training and application times Although a relatively simple approach, without prior understanding of human error the training time for TRACEr is likely to be quite high. The application time is also likely to be considerable due to the comprehensive nature of the method.

Reliability and validity There is limited reliability and validity evidence presented in the literature. Shorrock and Kirwan (2002) report encouraging preliminary reliability evidence; however, further reliability and validation evidence has not yet been forthcoming.

Tools needed TRACEr analyses can be undertaken using pen and paper along with the associated PEM, EEM, IEM, PSF taxonomies and the error detection and correction prompts.

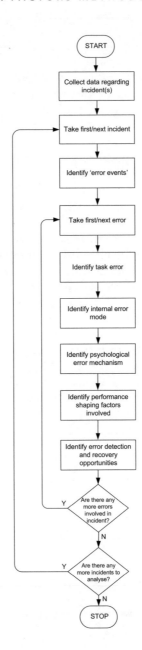

Flowchart 2-6   TRACEr

Recommended Reading

Isaac, A., Shorrick, S.T. and Kirwan, B. (2002) Human error in European air traffic management: the HERA project. *Reliability Engineering and System Safety*, 75, 257–72.

Shorrock, S.T. and Kirwan, B. (2002) Development and application of a human error identification tool for air traffic control. *Applied Ergonomics*, 33, 319–36.

Shorrock, S.T. (2006), Technique for the retrospective and predictive analysis of human error (TRACEr and TRACEr-lite). In: W. Karwowski (ed.), *International Encyclopedia of Ergonomics and Human Factors*, 2nd Edition, London: Taylor and Francis.

2.3.10 Event Analysis of Systemic Teamwork

Background and applications The Event Analysis of Systemic Teamwork framework (EAST; Stanton et al. 2005) provides an integrated suite of methods for analysing activity in collaborative systems that has recently been applied for accident analysis purposes (e.g. Rafferty et al. forthcoming). Underpinning the approach is the notion that collaborative activity can be meaningfully described via a 'network of networks' approach focusing on three different but interlinked perspectives: task, social and propositional networks underlying collaborative activity. Task networks represent a summary of the goals and subsequent tasks being performed within a system. Social networks analyse the organisation of the team and the communications taking place between the agents working in the team, and propositional networks describe the information and knowledge (distributed situation awareness) that the actors use and share in order to perform the teamwork activities in question. This 'network of networks' approach to understanding collaborative endeavour is represented in Figure 2-18 (adapted from Houghton et al, 2008).

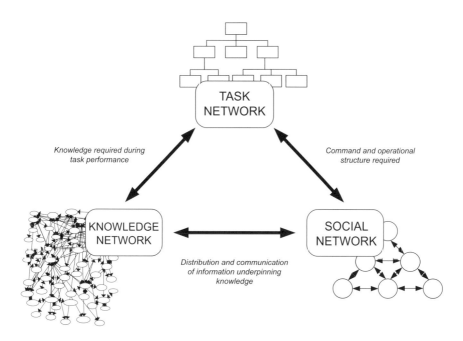

Figure 2-18 Network of networks approach to analysing collaborative activities; figure shows example representations of each network, including hierarchical task analysis (task network), social network analysis (social network) and propositional network (knowledge network) representations

To describe each of the three networks EAST uses a framework of different Human Factors methods; HTA (Annett et al. 1971) is typically used to construct task networks, SNA (Driskell and Mullen. 2004) is used to construct and analyse the social networks involved, and propositional networks (Salmon et al. 2009) are used to construct and analyse knowledge networks. In addition, Operation Sequence Diagrams (OSDs), Coordination Demands Analysis (CDA; Burke 2004), the CDM, and Communications Usage Diagram (Watts and Monk 2000) are used to evaluate various aspects of collaborative activity, including team cognition and decision-making, coordination and the technology used for communications.

Domain of application EAST is a generic approach that was developed specifically for the analysis of collaborative activities. To date, the approach has been used for this purpose in a number of different domains, including land warfare (Stanton, Salmon et al. 2010), airborne early warning and control (Stewart et al. 2008), naval warfare (Stanton et al. 2006), air traffic control (Walker et al. 2010), railway maintenance (Walker et al. 2006), energy distribution (Salmon et al. 2008) and emergency services (Houghton et al. 2006).

Accident analysis/investigation applications EAST has been applied for various purposes, including accident analysis. For example, Rafferty et al. (forthcoming) used EAST to analyse the Operation Provide Comfort Black Hawk helicopter friendly fire incident in which 26 personnel were killed when two US Army Black Hawk helicopters were mistakenly shot down by two US F-15 fighter jets. Component methods have also been applied for accident analysis; e.g. Griffin et al. (2010) used the propositional network approach for analysing the Kegworth British Midland Flight 92 air crash.

Procedure and advice Step 1: define analysis aims. First, the aims of the analysis should be clearly defined. This allows appropriate scenarios to be used and ensures that relevant data is collected. In addition, not all components of the EAST framework may be required, so it is important to clearly define the aims of the analysis to ensure that the appropriate EAST methods are applied.

Step 2: define task(s) under analysis. Next, the task(s) or scenario(s) under analysis should be clearly defined. This is dependent upon the aims of the analysis and may include a range of tasks or one task in particular. It is normally standard practice to develop an HTA for the task under analysis if sufficient data and SME access are available. This is useful later on in the analysis and is also enlightening, allowing analysts to gain an understanding of the task before any further observation and analysis begins.

Step 3: collect data regarding accident. Observational study is normally used as the primary data collection activity to support EAST analyses; however, this will often not be possible when using the approach for accident analysis purposes. Other useful data collection procedures involve interviews with SMEs or the personnel involved in the accident under analysis, documentation review (e.g. accident or inquiry report, standard operating procedures) and observations of the task(s) or system in question. When collecting data the following aspects of collaborative task performance are normally of interest: an incident timeline, including a description of the activity undertaken, the agents involved, any communications made between agents and the technology involved, the tasks performed and their purpose, the information used by different agents and teams, any tools, documents or instructions used to support activity, the outcomes of activities, any errors made and also any information that the analyst involved feels is relevant.

CHAPTER 2 · HUMAN FACTORS METHODS FOR ACCIDENT ANALYSIS

Step 4: conduct CDM interviews. Once the task under analysis is complete, or sufficient data regarding the task is collected, each 'key' agent (e.g. scenario commander, actors performing critical tasks) involved should be subjected to a CDM interview where possible. This involves dividing the scenario into key incident phases and then interviewing the agent involved in each key incident phase using a set of pre-defined CDM probes (e.g. O'Hare et al. 2000, see Chapter 2 for CDM method description).

Step 5: transcribe data. Once sufficient data is collected, it should be transcribed in order to make it compatible with the EAST analysis phase. An event transcript should be constructed. The transcript should describe the scenario over a timeline, including descriptions of activity, the actors involved, any communications made and the technology used. In order to ensure validity, the scenario transcript should be reviewed by one of the SMEs involved.

Step 6: re-iterate HTA and create task model. The data transcription process allows the analyst(s) to gain a deeper and more accurate understanding of the scenario under analysis. It also allows any discrepancies between the initial HTA scenario description and the actual activity observed to be resolved. Typically, collaborative activity does not run entirely according to protocol, and certain tasks may have been performed during the scenario that were not described in the initial HTA description. The analyst should compare the scenario transcript to the initial HTA, and add any changes as required. Once the HTA is complete, a task model depicting the main goals and tasks involved in the activity under analysis should be constructed.

Step 7: conduct co-ordination demands analysis. The CDA method involves extracting teamwork tasks from the HTA and rating them against the associated CDA taxonomy of teamwork behaviours: communication, situation awareness, decision-making, mission analysis, leadership, adaptability, assertiveness, and coordination. Each teamwork task is rated against each CDA behaviour on a scale of one (low) to three (high). Total co-ordination for each teamwork step can be derived by calculating the mean across the CDA behaviours. The mean total co-ordination figure for the scenario under analysis should also be calculated. It is possible for the analyst to conduct the CDA on their own using subjective judgement; however, it is often more useful if those involved or appropriate SMEs are used.

Step 8: construct comms usage diagram. Comms Usage Diagram (CUD; Watts and Monk 2000) is used to describe the communications between teams of agents dispersed across different geographical locations. A CUD output describes how and why communications between agents occur, which technology is involved in the communication, and the advantages and disadvantages associated with the technology used. A CUD analysis is typically based upon observational data of the task or scenario under analysis, although talk through analysis and interview data can also be used (Watts and Monk 2000). For accident analysis purposes it is useful to use CUD to identify any communications failures that played a role in the accident.

Step 9: conduct social network analysis. SNA is used to analyse the relationships between the agents involved in the scenario under analysis. It is normally useful to conduct a series of SNAs representing different phases of the task under analysis (using the incident phases defined during the CDM part of the analysis). It is recommended that the Agna SNA software package be used for the SNA phase of the EAST methodology. For accident analysis purposes it is useful to use SNA to identify any communications failures that played a role in the accident.

Step 10: construct operation sequence diagram. The OSD represents the activity undertaken during the scenario under analysis. The analyst should construct the OSD using the scenario transcript and the associated HTA as inputs. Once the initial OSD is completed, the analyst should then add the results of the CDA to each teamwork task step.

Step 11: construct propositional networks. The final step of the EAST analysis involves constructing propositional networks for each scenario phase. Following construction, information usage should be defined for each actor involved via shading of the information elements within the propositional networks.

Step 12: validate analysis outputs. Once the EAST analysis is complete, it is pertinent to validate the outputs using appropriate SMEs and recordings of the scenario under analysis. Any problems identified should be corrected at this point.

Advantages:

1. The analysis produced is comprehensive, and activities are analysed from various perspectives;
2. The framework approach allows methods to be chosen based on analysis requirements;
3. When used for accident analysis purposes, contextual factors, communications and situation awareness failures, task errors, and technological contributions are all considered;
4. EAST has been applied in a wide range of different domains for a range of purposes, including accident analysis (e.g. Rafferty et al. forthcoming);
5. The approach is generic and can be used for accident analysis purposes in any domain;
6. A number of Human Factors concepts are evaluated, including situation awareness, cognition, decision-making, teamwork, and communications; and
7. Uses structured and valid Human Factors methods and has a sound theoretical underpinning.

Disadvantages:

1. When undertaken in full, the EAST framework is a highly time-consuming approach;
2. The use of various methods ensures that the framework incurs a high training time;
3. In order to conduct an EAST analysis properly, a high level of access to the domain, task and SMEs is required;
4. When used for accident analysis purposes detailed data is required to support EAST analyses;
5. Some parts of the analysis can become overly time consuming and laborious to complete; and
6. Some of the outputs can be large, unwieldy and difficult to present in reports, papers and presentations.

Related methods The EAST framework comprises a range of different Human Factors methods, including observational study, HTA (Annett et al. 1971), CDA (Burke 2004), SNA (Driskell and Mullen 2004), the CDM (Klein et al. 1989), OSDs (Stanton et al. 2005), CUD (Watts and Monk 2000) and propositional networks (Salmon et al. 2009).

Approximate training and application times Due to the number of different methods involved, the training time associated with the EAST framework is high. Similarly, application time is typically high, although this is dependent upon the task under analysis and the methods used. Based on our experiences, once the data is collected it is not uncommon for EAST analyses to take up to and over one month to complete.

Reliability and validity Due to the number of different methods involved, the reliability and validity of the EAST method is difficult to assess. Some of the methods have high reliability and validity, such as SNA and CUD, whereas others may suffer from low levels of reliability and validity, such as the CDM and propositional networks approaches.

Tools needed Normally video and audio recording devices are used to record the activities under analysis. The WESTT software package (Houghton et al. 2008) supports the development of most EAST outputs. Various HTA software packages exist for supporting analysts in the construction of HTAs for the task under analysis. Also, Agna supports SNA and propositional network analyses. A drawing software package such as Microsoft Visio is also typically used to for the representational methods such as OSDs and CUD.

Example An example EAST analysis is presented in the final chapter.

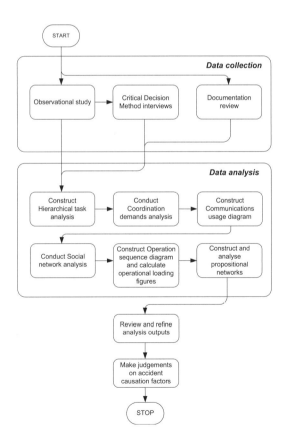

Flowchart 2-7   EAST

Recommended Reading

Stanton, N.A., Salmon, P.M., Walker, G., Baber, C. and Jenkins, D.P. (2005) *Human Factors Methods: A Practical Guide for Engineering and Design*. Aldershot, UK: Ashgate Publishing.

Walker. G.H., Gibson, H., Stanton, N.A., Baber, C., Salmon, P.M. and Green, D. (2006) Event analysis of systemic teamwork (EAST): a novel integration of ergonomics methods to analyse C4i activity. *Ergonomics*, 49:12–13, 1345–69.

3
AcciMap: Lyme Bay Sea Canoeing and Stockwell Mistaken Shooting Case Studies

3.1 ACCIMAP CASE STUDY 1: THE LYME BAY SEA CANOEING TRAGEDY

There is an acknowledged risk of both severe and frequent injury associated with active pursuits, especially those participated in for sport, active recreation or leisure (Finch et al. 2007; Flores et al. 2008). In Australia, injury-causing incidents are particularly problematic within the led outdoor activity domain (led outdoor activity is defined as facilitated or instructed activities within outdoor education and recreation settings that have a learning goal associated with them, including activities such as school and scout camping, hiking, harness sports, marine aquatic sports and wheel sports). Recent research indicates that the industry's understanding of such incidents is limited, and that the surveillance systems required to enhance it do not exist (Salmon, Williamson et al. 2010). As part of a programme of research, the overall aim of which is to develop an accident and injury surveillance system for the led outdoor activity domain in Australia, a range of accident analysis methods were applied to various led outdoor activity accidents. The aim of these analyses was to investigate different accident analysis frameworks for their utility in describing the accidents that occurred in this domain, and also to generate evidence for the systems perspective as an appropriate framework for accident causation in the led outdoor activity domain. The case study presented in this chapter describes the application of the AcciMap method for the analysis of a high-profile led outdoor activity incident, the Lyme Bay sea canoeing tragedy.

3.1.1 Incident Description

The Lyme Bay sea canoeing tragedy involved the death of four students whilst on an outdoor education activity trip in Lyme Bay, Dorset in the UK, on 22 March 1993. The activity involved a group of eight students, their schoolteacher, a junior instructor and a senior instructor engaging in an introductory open sea canoeing activity in Lyme Bay. After entering the water, the schoolteacher capsized repeatedly close to shore. Whilst the senior instructor attempted to right the schoolteacher, the junior instructor and eight students became separated from them and were blown out to sea. Whilst out at sea, high wind and wave conditions, lack of appropriate equipment, and inexperience led to all of the canoes being swamped and subsequently capsizing. As a result, the eight students and the junior instructor were left in the water with all canoes abandoned. After a delayed response and rescue attempt four students lost their lives through drowning.

3.1.2 Data Sources and Data Collection

The official inquiry report (Jenkins and Jenkinson 1993) was used as the primary data source for the AcciMap analysis. Other sources of information were also reviewed, including newspaper reports, textbooks, and journal articles describing the incident.

3.1.3 Analysis Procedure and Resources Invested

One Human Factors researcher with significant experience in the application of the AcciMap method, and other accident analysis approaches, performed the analysis based on the data contained in the official inquiry report (Jenkins and Jenkinson 1993). In order to ensure reliability and validity, three Human Factors researchers independently reviewed the analysis and input data and any subsequent disagreements were resolved through further discussion and consensus. Following this, an expert panel of outdoor education and adventure activity providers reviewed and refined the analysis outputs (in consultation with all four Human Factors researchers).

3.1.4 Outputs

The AcciMap for the Lyme Bay incident is presented in Figure 3-1.

3.1.5 Discussion

The aim of this analysis was to test the AcciMap method as an approach for analysing accidents occurring in the led outdoor activity domain. Specifically the authors wished to identify an accident analysis method that could identify contributory factors across the entire led outdoor activity system. A secondary aim of the analysis was to test Rasmussen's risk management framework in the led outdoor activity domain (see Salmon, Williamson et al. 2010). The analysis identified failures at each of the six organisational levels specified by the AcciMap approach. A summary of the factors identified at each level is given below.

Equipment and surroundings A failure by the activity centre and instructors to provide appropriate equipment for the canoeing activity was cited by the inquiry report as a key contributing factor to the accident. For example, spray decks, the device which creates a watertight seal at the point where the user sits in the canoe, were not present in the students' canoes, despite the students all being inexperienced in the operation of canoes and also in sea canoeing in general (the instructors used spray decks). This was identified as a key factor in the ease with which the canoes were swamped and capsized. Further,

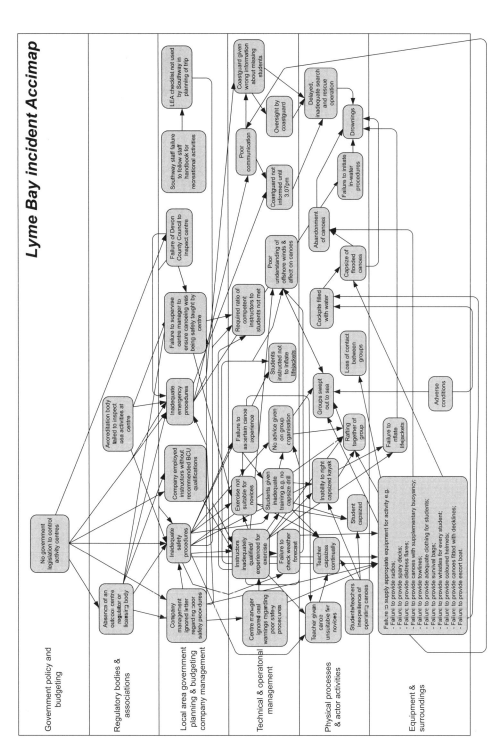

Figure 3-1   Lyme Bay tragedy AcciMap

the students' canoes did not have supplementary buoyancy (which is recommended for sea canoeing) and were not fitted with deck lines, which meant that they could not be tied to each other or towed in the event of the group becoming separated. It was also concluded that the students were inadequately clothed for sea canoeing activities, and the lifejackets worn by the students were not inflated, on instruction from the instructors, and had no whistles attached to them.

Various other pieces equipment that would normally be used on beginner sea canoeing exercises was not provided. This included radios, distress flares, towlines and a survival bag. Also, the school teacher who capsized initially, an inexperienced sea canoeist, was given a Laser 350 canoe (Jenkins and Jenkinson 1993), which was shorter and lighter than the other canoes and more suitable for use by a more experienced canoeist. Finally, at this level the environmental conditions also played a role in the accident. Although conditions on the shore were not overly adverse, the offshore winds and high seas were key factors in sweeping the students out to sea and also capsizing the students whilst out at sea.

Physical processes and actor activities At the physical processes and actor activities level, most of the contributing factors identified relate to the instructors' and students' inability to respond to the unfolding situation, which was a function of their lack of experience and qualifications for such an activity and the absence of appropriate safety and emergency procedures. The trigger events were the continued capsizes of the teacher's canoe, and the collective inability of the senior instructor and teacher to right the capsized canoe. Whilst the senior instructor attempted to right the capsized canoe, the junior instructor proceeded to raft together the students (i.e. organize tightly together in formation), although prior to this no advice had been given on group organization. Influenced by the offshore winds, the rafted group was quickly swept out to sea, and, due to the absence of distress flares and radios, the two groups (instructor/teacher group and rafted group, including eight students and one junior instructor) lost contact with one another. Since the rafted group could not be tied together (due to the absence of tow/deck lines), the high wave conditions out at sea, coupled with the lack of spray decks, led to multiple capsizes in the junior instructor/student group. As a result the canoes became swamped with water until eventually all were submerged with only one upturned canoe to hold on to. After a failed attempt to paddle ashore using the upturned canoe, the instructor and students abandoned the canoe. Whilst in the water, standard in-water procedures, designed to keep submerged entities warm and prevent hyperthermia, were not initiated.

The emergency response to the unfolding situation is also important. The alarm was not raised on-shore until over three hours after the group should have returned. Further, the coastguard was not immediately notified of the missing group by the site manager; rather, he spent some time searching the shoreline for the missing canoeists. Finally, the coastguard was wrongly informed that the instructors were well qualified, over 18 years of age, and were well equipped for the activity, all of which was not the case. Once the rescue was underway, the coastguard themselves made a number of errors, although this was not the focus of the analysis presented.

Technical and operational management At the technical and operational management level, decisions and actions both preceding the accident and on the day of the tragedy played a significant role in the failures at the lower levels. Before the incident, the activity centre's manager (and also the company management) failed to heed the content of a letter sent by two previous employees to management regarding unsafe practices, poor safety procedures, and the provision of inadequate equipment for the sea canoeing activities taught at the centre. The centre manager also failed to heed verbal warnings from the same two employees regarding the fact that canoeing was not being safely taught at the centre (Jenkins and Jenkinson 1993). The competence, qualifications, and expertise of the instructors involved was also problematic, with the instructors employed by the centre not being sufficiently qualified for sea canoeing, which represents a failure on the part of the manager to employ adequately qualified staff. The exercise itself was not suitable for novice canoeists; further, the inquiry report suggests that no attempt was made by staff to ascertain the experience levels of the students or teacher. As a result, the training given to the teacher and students prior to the trip in the centre's swimming pool was inadequate, involving no capsize drills designed to teach procedures to right and re-board canoes.

On the day of the tragedy, neither of the two instructors leading the activity checked the weather forecast, and no attempt was made to ascertain the conditions out at sea. Students were also instructed to not inflate their lifejackets. Due to the instructors' lack of qualifications and the type of activity being undertaken, the required ratio of competent instructors to students was also not met.

Local area government planning and budgeting, company management At the company management level, the company's management also failed to heed the content of the letter sent by two previous employees regarding the safety of canoeing activities at the centre in question. The managing director was also charged with failing to devise and enforce safe procedures for executing sea canoeing activities, and the emergency procedures in place were also found to be inadequate. The employment of inadequately qualified instructors is also represented at this level, with the company management failing to procure staff who were suitably qualified to provide safe canoeing activities. There was also a failure on the part of the company's management to supervise and ensure that safe activities were being provided at the centre in question. Finally, the inquiry report also concluded that the school involved had not planned the trip adequately in failing to completely follow their own staff handbook for organizing such trips, and a local education authority checklist for planning such activities was not used.

Regulatory bodies and associations At the time of the incident, there was no regulatory or licensing body for outdoor activity centres. This meant that unsafe practices and procedures could, to a large extent, continue unchecked without reprisals. Following the incident, the inquiry report recommended that a national independent system of registration and regulation of outdoor activity centres be developed as a matter of urgency, and also that no school or youth group should be permitted to use such centres until they had been approved for registration by an appropriate body. The absence of a regulator and legislation allowed the company to continue with unsafe and inappropriate procedures, and to employ inadequately qualified staff. The report concluded that, if a regulatory body had been in place at the time, the concerns reported by previous employees would undoubtedly have been reported to the body, and appropriate action would have ensued.

In addition, although the centre in question was accredited by the British Activity Holidays Association, the inquiry report suggested that no examination of sea canoeing activities was undertaken, rather, only land and pool-based activities were examined.

Government policy and budgeting At the government policy and budgeting level, the absence of legislation to control outdoor activity centres was identified as a key factor in many of the lower level failures. This meant that there was no regulating or licensing body overseeing outdoor activity providers, which enabled the centre to continue engaging in unsafe practices, despite these being identified and documented by previous employees of the centre.

3.1.6 Summary of Key Findings

The analysis found a number of key failures, all of which contributed to the Lyme Bay sea canoeing tragedy. These failures were related to the equipment used, the instructors leading the activity, the activity centre and company management, and government legislation at the time. The findings therefore demonstrate that contributory factors were present across all levels of the led outdoor activity system, which not only provides support for the use of AcciMap in this domain but is also consistent with the systems perspective on accident causation. These causal factors were also spread across various different actors and organizations, including the instructors, the led outdoor activity provider, the local authority, and government. This is also consistent with the systems approach to accident causation. The analysis also demonstrated the utility of the AcciMap approach for analysing led outdoor activity accidents. Provided sufficient data is available, the method is capable of identifying failures across all levels of organisational systems, is simple to apply, and the output is easily interpretable. The output also supports the development of appropriate systems-focused countermeasures and remedial strategies, as opposed to individual-oriented measures that ignore the wider systemic causal factors often produced from individual 'operator at the sharp end' accident analyses (Dekker 2002; Reason 1997).

3.1.7 Acknowledgement

The research described in this case study was funded by the Department of Planning and Community Development (Sport and Recreation Victoria), along with contributions from various led outdoor activity providers. The authors also wish to acknowledge the members of the project steering committee, who provided initial data for the analysis and also validated and subsequently formed the expert panel for refining the analysis presented. Finally, the authors would like to acknowledge that the description presented is derived from the following journal article:

Salmon, P.M., Williamson, A., Lenné, M.G., Mitsopoulos, E. and Rudin-Brown, C.M. (2010) Systems-based accident analysis in the led outdoor activity domain: application and evaluation of a risk management framework. *Ergonomics*, 53:8, 927–39.

3.2 ACCIMAP CASE STUDY 2: THE STOCKWELL JEAN CHARLES DE MENEZES SHOOTING

On 22 July 2005 in South London, UK, two members of the Metropolitan Police Service's (MPS) specialist firearms department entered Stockwell underground station with orders to 'stop' a suspected suicide-bomber. Upon entering the platform, the officers were directed towards a possible terrorist suspect, who had been followed from his home to Stockwell station by surveillance officers. The firearms officers approached the man and between them fired eight shots at close range. This man was later found to be Jean Charles de Menezes (JCdM), an innocent Brazilian national, working in London as an electrician.

The AcciMap method was used by the authors to analyse the Stockwell shooting. The aim of the analysis was to identify both system failures at a component level, as well as the overarching bureaucratic or political issues that contributed to the incident. The analysis also formed part of an overall research effort focusing on the use of formative modelling approaches for describing activities in complex sociotechnical systems (see Jenkins et al. 2010).

3.2.1 Incident Description

In July 2005, the city of London was faced with the reality of unprecedented suicide bombing attacks. On 7 July, four bombs were detonated by suicide bombers at different locations throughout London's transport network (three on underground trains at Russell Square, Aldgate, Edgware Road and one on a bus at Tavistock Place) killing 52 innocent people, along with the four suicide bombers. Hundreds more were injured, and significant damage was caused to buildings and infrastructure. The impact of this terrorist act on the citizens of London, and indeed the world, cannot be overstated. In addition to the shock, grief and suffering, there was a heightened fear of further terrorist attacks. Based upon credible intelligence, Britain was placed on high alert, indicating that further attacks were imminent.

Two weeks after the 7 July bombings, on the afternoon of 21 July 2005, fears of further attacks were confirmed when evidence of four failed explosive devices was found at various locations around London. All four of the bombers escaped, and the subsequent examination of the devices recovered linked them to those used on 7 July. Further evidence, a gym membership card, was found within a bag containing one of the unexploded devices. The membership card linked one Hussain Osman to the crime and his address was subsequently traced to 21 Scotia Road – an address also used by another suspect of interest to the Metropolitan Police. Based on a comparison of images obtained from CCTV cameras at the scene of the failed explosions with images of Osman, it was judged that there was a 'good likeness' between the two (IPCC 2007, 20).

This led to the initiation of a proactive operation, orchestrated from the central police control room, known as 'Room 1600'. The Gold Commander decided that an operation should be mounted the following morning (22 July) to apprehend Hussain Osman at Scotia Road. The aim of the operation, as defined by the Gold Commander, was to control the identified premises through covert surveillance, and follow any person leaving the premises until it was deemed safe to challenge them. According to briefings (IPCC 2007), if the stops identified other residents of the flats, then any intelligence opportunity would be maximised.

Based upon intelligence gathered, the officers involved were briefed that they might encounter suicide-bombers. Whilst the attacks on 7 July were the first instances of suicide-bomb attacks on the UK, the situation had been anticipated and the Metropolitan Police Service had developed a number of policies to respond to this form of threat. According to the IPCC report (2007), a strategy had been developed, known as: 'Operation Kratos'. According to the Metropolitan Police Service (MPS 2008) 'Operation Kratos is the overarching operational name given to the national police response to protect the public from the threat posed by suicide-bombers, either on foot or in a vehicle. The use of potentially lethal force may be a police officer's only option, where the information and intelligence indicates that a person is in possession of an explosive device, and whose only intent is to kill and maim as many people as possible. Each use of force by police must only ever be used when proportionate to the danger and when absolutely necessary'. According to the IPCC (2007), as the suicide-bomber intends to die in their attack, and as many explosive devices are extremely sensitive to impact, a unique approach has to be taken. The tactics employed under Kratos are to deliver a fatal shot to the head, before any attempt to negotiate, which provides the terrorist with opportunity to detonate (although the MPS are keen to stress that Kratos is not a 'shoot-to-kill policy', rather a policy to incapacitate). The Kratos policy is also explicit in stating that arrest at an address or other location at an earlier stage will always be a preferred option. According to IPCC (2007), Kratos operations are controlled by a Designated Senior Officer (DSO) who, based upon sufficient information to justify use of lethal force, makes the decision to shoot. A small cadre of officers at Commander Rank within the Metropolitan Police perform the role of Designated Senior Officer (DSO). On 22 July, Commander Cressida Dick was appointed to undertake the role of DSO.

The DSO used a number of specialist resources available to her, including the Metropolitan Police Service's specialist firearms department (CO19), Special Branch (SO12, an investigative unit dealing with national security matters), and the anti-terrorist branch (SO13). On 22 July, Special Branch were given the responsibility of observing those leaving Scotia Road with the aim of collecting intelligence and ultimately identifying Osman. The anti-terrorist Branch were to be responsible for assisting with the arrest of the suspects and de-briefing anyone coming out of the flats, and the firearms team was responsible for challenging armed suspects. The firearms team was trained to respond to incidents covered by 'Operation Kratos'. Although some Special Branch officers (SO12) carried firearms on 22 July, these were for their own protection and for the protection of the public. According to IPCC (2007), the training given to the Special Branch officers does not enable them to be used as a resource to arrest armed suspects. However, evidence was given at the Central Criminal Court that armed officers from the Special Branch would have been used as a last resort.

The operation was split between the control room (Room 1600 at police headquarters, New Scotland Yard) and various posts on the ground. Policy dictates that the DSO, (Commander Cressida Dick, based in Room 1600), controls the operation. She was assisted by between 20 and 30 support staff in Room 1600, including tactical advisors and a number of dedicated liaison officers. Information was communicated to the control room via 'party' radio links and 'open' peer-to-peer telephone lines. On the ground, there were three separately located teams; the Special Branch (SO12) observation team, the anti-terrorist branch (SO13) debrief team and the CO19 firearms team. As well as coordinating with the control room, the three teams were directed by the ground Silver Commander 'DCI C'. While bi-directional communications existed between the control room and officers on the ground, much of the communication flow relating to the identification of the suspect was informed by listening in to the Special Branch radio network.

In accordance with the direction provided by the Gold Commander, undercover surveillance officers from Special Branch, briefed on the unfolding situation, were deployed at Scotia Road to identify all individuals leaving the communal door to the block of flats. They were first alerted to JCdM at 09:33hrs on 22 July 2005, when he left by the communal door. They followed JCdM for 33 minutes, in which time he walked to catch a bus at a nearby bus stop, boarding a bus towards Brixton. At Brixton, JCdM alighted and walked towards Brixton tube station. Upon realising that the station was closed, he ran back to the same bus, and continued to Stockwell underground station. Once at the station, JCdM calmly entered the tube, stopping to collect a free newspaper. He then passed through the ticket barriers using his pre-paid 'oyster-card', walked down the escalators and boarded a stationary train. Within moments, two CO19 officers arrived and, aiming at his head, shot and killed him.

3.2.2 Data Sources and Data Collection

The primary source of data used for the AcciMap analysis was the detailed IPCC report (IPCC 2007); however, the analysis was supplemented by other data sources, including news reports surrounding the incident and a BBC Panorama documentary detailing the activities leading up to the incident.

3.2.3 Analysis Procedure and Resources Invested

One Human Factors researcher with significant experience in the AcciMap method conducted an AcciMap analysis of the JCdM shooting incident. This was then independently reviewed and validated by three Human Factors researchers. Any outstanding disagreements were discussed between all four researchers until consensus was reached.

3.2.4 Outputs

The AcciMap for the Stockwell shooting incident is presented in Figure 3-2. The events presented within the AcciMap are coded based on when, within the overall chain of events, they occurred. The coding categories used include:

1. Pre operation. Includes those events that occurred prior to the operation commencing;
2. Pre JCdM leaving. Includes those events that occurred once the operation had commenced but before JCdM left his Scotia Road residence;
3. Scotia Road to bus. Includes those events that occurred whilst JCdM was making his way to the bus at Tulse Hill;
4. Bus to tube station. Includes the entire bus journey taken by JCdM, from Tulse Hill to Brixton, and then Brixton to Stockwell;

5. In tube station. Includes events from the moment JCdM entered the station to the point at which he was shot.
6. All situations. The All Situations category is used to capture poor coordination, which was found to affect all events.

As represented within the AcciMap, failures that contributed to the JCdM shooting were identified at each of the six levels. A discussion of the findings is presented below.

3.2.5 Discussion

Equipment and surroundings Various failings were identified at the equipment and surroundings level. Lack of equipment, or failure to set up available equipment, affected the ability to capture an image of JCdM throughout his journey. The obvious opportunity to gain a clear unobstructed view of all residents leaving the flat at Scotia Road was from an observation van manned by surveillance officer, 'Frank'. This opportunity was not utilised effectively for a number of reasons. Firstly, the surveillance officer chose to switch the video camera on only when needed, switching it off when not in use to conserve battery power, rather than connecting it to the van's power supply with the available cables and leaving it switched on constantly. As a result, he was only able to capture video footage of six of the eight people who left the building. As JCdM left the flat, Frank was urinating into a plastic container, and was therefore unable to switch on the video camera. Additional technical constraints meant that even if the surveillance officer had been able to capture an image of JCdM, he would have been unable to relay the image. A 'smart phone' or laptop with mobile Internet would have been sufficient to allow Frank to distribute images to other surveillance officers and the command team.

Some of the communication breakdowns identified by the IPCC report (2007) can be partially attributed to the technology available to those involved. The radio of one of the special branch officers (Ken) malfunctioned at a critical moment shortly before JCdM boarded the bus for the first time. The officer was unable to transmit or hear what was being communicated and was therefore unable to get into a position to observe JCdM's face. Failures in the radio specification and infrastructure, first identified as far back as 1987 during the aftermath of the Kings Cross underground station fire, also meant that officers were unable to communicate while underground in the tube station.

The environment in the central control room (Room 1600) was criticised by some of the officers in their statements to the IPCC. The control room was described as 'very noisy', and it was suggested that it was 'necessary to shout to ensure senior officers were aware of what was going on'. This may have had an impact on the passage of information (the dotted line in Figure 3-2 indicates a possible, contentious, link).

Physical processes and actor activities Failures to challenge JCdM at a number of points on his journey to the tube station can be directly linked to operational deployment decisions which are discussed in the technical and operational management section below. Nearly all of the failures identified at the physical processes and actor activities level have precursors of latent conditions, indicating that either equipment failures, or previous decisions, commands, or policies had a significant impact on the action taken. The exceptions to this are the DSO, Commander Dick, arriving 25 minutes late for Commander McDowall's briefing, which meant she was unable to influence the firearms

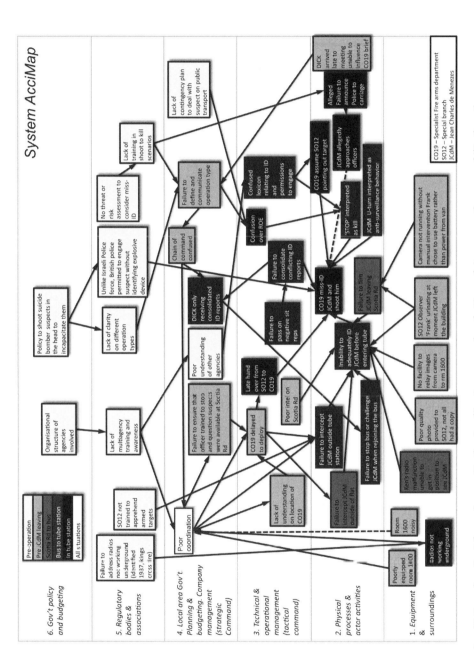

Figure 3-2 AcciMap (coded by phase, links indicate causal relationships, weak causal links shown as dotted lines)

team briefing, and officers mistaking JCdM's returning to the bus, after realising that Brixton station was closed, as anti-surveillance behaviour. These can be considered to be active failures that were not directly influenced by latent conditions (as defined by the system boundary). Salient failures at this level include the misinterpretation of the 'Stop' command issued by the DSO, and the meaning of officers pointing out the suspect, and the failure to get into a position to identify JCdM.

Technical and operational management The technical operational management level captures the decision and omissions made at a tactical level. These can be broken down into two broad categories: communication and coordination. The main coordination error resulted in the firearms team being unable to challenge JCdM until the final moments of his 33-minute journey. They failed to arrive at Scotia Road in time, and were unavailable to challenge JCdM as he left the bus in Brixton and in Stockwell. The requirement to have firearms officers in place to challenge suspects leaving Scotia Road was established at 04:55hrs. At 08:33hrs, one of the surveillance team leaders expressed concern over the distance between Scotia Road, and the current firearms team location (IPCC 2007, 55). The initial instructions of the Gold Commander (who had no involvement in the operation after issuing his orders) were, 'to control the premises at Scotia Road through covert surveillance, follow any person leaving the premises until it was felt safe to challenge them and then stop them' (IPCC 2007, 24). According to the IPCC report (2007, 122), however, there was a failure of this policy. There was no adequate effort to put police resources in place at Scotia Road to give effect to the policy. As a result, not one of the eight people who left the flat before JCdM did was stopped (the policy was to gain intelligence from questioning these people).

Communication can be defined as 'the process by which information is clearly and accurately exchanged between two or more team members in the prescribed manner and with proper terminology; the ability to clarify or acknowledge receipt of information' (Cannon-Bowers et al. 1995, 345). In this case, communication errors led to confused accounts of the location of the firearms officers. There were also communication errors in the consolidation of a number of sightings. There was a failure to properly consider negative sightings, and a tendency to over-emphasise positive identifications, resulting in a translation of a 'possible' to a 'positive' identification. There appears to be confusion surrounding the Rule of Engagement (RoE). This is further complicated by the use of an ambiguous lexicon, that is, using terms in commands such as, 'Stop him', which are open to interpretation. As exposed in previous studies of multi-agency work (Salmon et al. 2009), the use of a common language is integral to multi-agency or team command and control.

Local area government planning and budgeting; company management The company management level captures the decisions and omissions at the strategic level. A number of failings were identified in the running of the central control room (Room 1600). At the 07:45hrs firearms team briefing, the officers were informed that the instruction for a critical shot would come direct from the DSO. Based upon this responsibility placed on the DSO, it is imperative they are able to maintain an accurate and up-to-date picture of the operation on the ground. It is also imperative that they are able to issue commands that are correctly acted upon. The strategic order of preventing all persons leaving the Scotia Road address was clearly not followed. The command and control function failed to ensure that key assets were in the right place at the right time. This resulted in a failure to prevent the suspect from entering public transport. No attempt was made to challenge JCdM during his five-minute walk to the bus stop, nor when he got off the bus at Brixton, nor was he prevented from entering the tube network (both buses and tube trains had been previous targets on 7 July and 21 July).

Failures in communications into the control room led to the DSO receiving poorly consolidated summaries of the identification attempts. Whilst these communication errors can be partially attributed to technical shortfalls, they can also be attributed to failure in the command process and data fusion. Failures in communications out of the control room include the lack of clear, unambiguous commands.

Regulatory bodies and associations In this instance, the regulatory bodies level has been used to describe the policies and the organisational structure of the Metropolitan Police Service. Shortfalls in policy and organisational structure undoubtedly contributed to the breakdowns in coordination as well as poor understanding of other organisations' interpretation of the lexicon. In an attempt to safeguard operations involving suicide-bombers, a risk-assessment was conducted on the previous day, by officers from the firearms team. However, in accordance with the policy at the time, this assessment did not consider the risk of misidentification or uncertainty regarding the identification of a suspect. Nor did the assessment consider a suspect leaving the premises before firearms resources were in place.

One of the key findings from the AcciMap analysis, not explicitly identified in the IPCC report (2007), relates to the organisational structure of the MPS during terrorist operations. The functional divisions between Special Branch and the firearms team are a cause for concern in fast-paced operations where a reliable positive identification has not been possible. Assigning the responsibility of surveillance to one organisation (Special Branch; SO12), and the responsibility of 'controlling' the suspect to another (the firearms team; CO19), requires the communication of complex situational understanding from one team to the other. In terms of the understanding of the suspect, and the likelihood of him being a suicide-bomber, the surveillance officers were clearly in a much better position to make a final call. The system failed to support a process for consolidating this understanding in such a way that it could be meaningfully transmitted to the command team or the firearms officers. Based upon the findings of this analysis, it is contended that the same individuals or team should conduct both surveillance and control of suspects' roles in future similar operations.

Government policy and budgeting The policy to shoot a suspect in the head without visually identifying an explosive device or weapon undoubtedly had an impact on the shooting. A safeguard, such as that required in Israel, would have forced the officers to 'control' the suspect until they were certain he was carrying a device. Clearly, the aim of the critical-shot policy is to preserve human life by preventing a suicide-bomber from detonating a device. However, a careful balance has to be made between the preservation of innocent misidentified individuals as well as a suicide-bomber who can be controlled and apprehended safely.

3.2.6 Summary of Key Findings

The key findings and recommendations reported by the IPCC (2007) are represented in the AcciMap. By representing these findings in a graphical format, it was possible to link the conditions and actions that led to the incident. Furthermore, the AcciMap analysis revealed additional insights and recommendations not explicitly described in the IPCC (2007) report. As identified at the Regulatory bodies and associations level of the AcciMap, the organisational structure of the MPS may need to be reconsidered in response to new threats and conditions. It is contended that the current structure of dedicated surveillance officers (sensors), distributed commanders, and dedicated shooters, may not be optimal for all operations captured under the umbrella of Kratos. A less rigid, less bureaucratic system would be able to adjust dynamically to the operational context, e.g. in allowing the Special Branch officers to step in and challenge JCdM at an earlier juncture. Whilst this recommendation for a more flexible system may be new to this incident, it has frequently been found in other studies of complex sociotechnical systems analysed using AcciMap. According to Rasmussen (1997), the pace of change will often outstrip changes in management structures, legislation and regulations. People are constantly facing situations not covered by available rules and procedures. The humans within a system are required to adapt current working practices in real time to fit the changing situation. As these environmental changes are often unpredictable, systems need to be developed with appropriate flexibility to adapt. As Woo and Vicente (2003) point out, risk management continues to be somewhat of a hit and miss affair because each large-scale accident appears to be unique. Analyses of high-profile disasters typically reveal an idiosyncratic set of factors, which are never likely to be repeated in the same system, let alone in other systems in the same sector or in other sectors. However, it is contended that changes made at higher levels of the AcciMap are more likely to be applicable to a wider range of situations.

3.2.7 Acknowledgement

The authors would like to acknowledge that the description presented above is derived from the following journal article:

Jenkins, D.P., Salmon, P.M., Stanton, N.A. and Walker, G.H. (2010) A systemic approach to accident analysis: a case study of the Stockwell shooting. *Ergonomics*, 3:1, 1–17.

4

The Human Factors Analysis and Classification System: Australian General Aviation and Mining Case Studies

4.1 HFACS ANALYSIS OF AUSTRALIAN GENERAL AVIATION ACCIDENTS

4.1.1 Introduction

The first case study involves an application of HFACS to the analysis of Australian General Aviation (GA) incidents. The analysis presented was undertaken as part of an overall programme of research that involved the analysis of Australian GA incident data held by aviation insurers for trend identification and reporting. At the time of the study, initial research revealed that there were various inconsistencies in the manner in which GA accident and incident data were collected, classified, stored, analysed, and reported. Notably, a review of existing databases indicated that they did not contain sufficient information to support comprehensive analysis of the causal factors involved in GA accidents and safety compromising incidents (Lenné et al. 2007).

As part of the latter phases of the research programme, existing GA data held by insurers was analysed using the HFACS approach. The purpose of this was to identify accident and incident trends, to validate the HFACS approach as a suitable one for GA accident data collection and analysis purposes, and also to inform the development of an accident and incident data collection template and training package for GA insurance assessors.

4.1.2 Incident Description

Sequential claims from three aviation insurers were sampled, covering the time period 25 February 2002 to 13 July 2004. These claims were obtained from the Victorian branches of the insurers, which receive claims from Victoria, New South Wales, and Queensland. Following removal of non-applicable cases (e.g. cases of malicious damage), a total of 188 GA incidents were deemed eligible for the HFACS analysis component. This represented

68 per cent of all insurance claims examined over this period by the insurers involved. General characteristics of the incidents are discussed below.

Aircraft type and purpose of flight   The type of aircraft involved was specified in 89 per cent of the incidents. A summary of the aircraft types involved is presented in Table 4-1.

The primary reason why the flight was initiated was recorded for 95 per cent of the cases. A summary of flight purpose is presented in Table 4-2.

As shown in Table 4-2, private flights made up 37 per cent of the incidents analysed, followed by flights for agricultural purposes and flight training (each 13 per cent) and charter flights (12 per cent). Other flights involved included aerial work, scheduled flights, and business flights.

Table 4-1   Aircraft types involved in incidents analysed

Aircraft type	No. of incidents	% of total
Fixed-wing single engine	99	52.7
Fixed-wing multi-engine	50	26.6
Rotor wing	17	9.0
Ultra light	1	0.5
Missing	21	11.2
Total	188	100

Table 4-2   Summary of flight purpose

Flight purpose	No. of incidents	% of total
Private	69	36.7
Agricultural	25	13.3
Flying training – Solo – Dual	25 19 6	13.3
Charter	23	12.2
Aerial work	19	10.1
Scheduled flight	15	8.0
Business	2	1.1
Other/missing	10	5.3
Total	188	100

CHAPTER 4 • THE HUMAN FACTORS ANALYSIS AND CLASSIFICATION SYSTEM

Pilot age, gender and experience The age of the pilot was recorded for one quarter of the incidents analysed (n = 48). The age of pilots involved ranged from 22 to 79 years, with an average age of 42.3 years. The gender of the pilot was known for 79.3 per cent of incidents and the male:female ratio was 97:3.

A range of data were recorded regarding each pilot's experience, including the pilot's total flying time, flying time on the type of aircraft involved in the incident, pilot's total flying time in the last 90 days, and flying time on the type of aircraft involved in the incident in the last 90 days.

Total pilot flying time ranged from 40 to 20,000 hours with an average of 3,558 hours. Total pilot flying time for the type of aircraft involved in the incident ranged from two to 9,855 hours, with an average of 859 hours. Pilot flying time in the 90 days prior to the incident ranged from two to 1,000 hours with an average of 78 hours. Pilot flying time in the last 90 days on the type of aircraft involved in the incident ranged from zero to 238 hours, with an average of 57 hours.

Pilots of fixed-wing multi-engine aircraft were reported to have the highest average flying hours in total and in the previous 90 days, however, rotor wing pilots had a higher number of average hours on that type of aircraft both overall and in the last 90 days. Pilots of scheduled flights and charter services had the highest average total flying hours, however, pilots in the combined aerial work and business category had the highest average number of hours on their aircraft type. Those pilots involved in flight training recorded the highest average number of hours' flight in the previous 90 days and pilots of scheduled flights the highest average in the 90 days on that particular type of aircraft.

Of the pilots involved, the breakdown by licence type was as follows: commercial (34 per cent), private (20 per cent), air transport (15 per cent), student (2 per cent), with another 29 per cent missing.

Phase of operation, outcome, and effect on flight The incidents were initially classified via three factors: phase of operation, outcome, and effect on flight. For example, a lightning strike while cruising that had no effect on flight was coded as cruise-lightning strike – no effect on flight. A summary of the incidents analysed by phase of operation is presented in Table 4-3.

Landing was the most common phase of operation represented in the data set, with 29 per cent of all incidents occurring during landing, followed by 15 per cent during taxiing, 13 per cent during takeoff and 12 per cent en route or cruising.

A summary of the outcomes of the incidents analysed is presented in Table 4-4.

A total of 238 outcomes were reported over the 188 incidents. The most common outcomes were propeller strikes (19 per cent), impact with the ground – not collision (16 per cent), collision with an object on the ground (14 per cent), and heavy landing (13 per cent).

The effect on the flight was specified in 137 incidents, however this was applicable in only 122 incidents. A summary of the effect on flight of the incidents analysed is presented in Table 4-5.

The most commonly reported effect on flight was the aircraft coming to rest/the travel ended (24 per cent), e.g. landing off the side of the runway or prop strike on taxiing. This was followed by a forced landing (13.8 per cent).

Table 4-3   Phase of operation

Phase of operation	No. of incidents	% of total
Hangared	3	1.6
Standing	3	1.6
Start-up	6	3.2
Taxiing	28	14.9
Take-off	24	12.8
Climb	4	2.1
En-route/cruise	22	11.7
Manoeuvring	8	4.3
Spray-run	5	2.6
Aerobatics	3	1.6
Descent	8	4.3
Approach	2	1.1
Landing	54	28.7
Maintenance	8	4.3
Towing/parking	8	4.3
Other/missing	2	1.1
Total	188	100

Table 4-4   Outcome of incident

Outcome of incident	No. of incidents	% of total
Wire strike	6	2.5
Animal/bird strike	11	4.6
Lightning strike	8	3.4
Hail damage	5	2.1
Gear up landing	12	5.0
Heavy landing	31	13.0
Overshoot runway	8	3.4
Side run off from runway	9	3.8
Landing off runway	8	3.4
Propeller strike	46	19.3
Heavy impact with ground (not collision)	37	15.5
Collision with an aircraft (on ground and in air)	6	2.5
Collision with terrain	11	4.6
Collision with object on ground	33	13.9
Fire	7	2.9
Total	238	99.9

Table 4-5 Effect on flight

Effect on flight	No. of incidents	% of total
None	24	12.8
Rejected take-off	8	4.2
Diversion	4	2.1
Precautionary landing	14	7.4
Engine shutdown	1	0.5
Forced landing	26	13.8
Aircraft came to rest/end of journey	45	23.9
Not applicable	15	8.0
Other/missing	51	27.1
Total	188	

4.1.3 Data Sources and Data Collection

As stated above, the data was derived from the insurance claims held by the Victorian branch of the aviation insurer involved covering the period 25 February 2002 to 13 July 2004.

Extraction of insurance data   The key components of each claim file were the insurance claim form and the assessor report, lodged by the insurer-appointed loss adjustor. The insurance claim form is completed by the assured and submitted to the insurer. The insurer then determines whether or not an assessor will be appointed to investigate the claim.

A database was developed to store the de-identified data extracted from the assessor reports. The database was structured such that two levels of data could be stored. Firstly, the data extracted were those items typically documented by the insurance assessors, and covered the broad headings outlined below. These data were entered by the research team. The second level of data collected were the assessments of each claim made by the expert panel members using HFACS. This is discussed in more detail later.

Where available, data items extracted from the files were:

- Claim reference number
- Aircraft type
- Aircraft registration number
- Brief incident description
- Location of incident
- Date of incident
- Synopsis
- Introduction
- Circumstances
- Damage sustained
- Year of aircraft manufacture
- Total air hours for aircraft

- Gender of pilot
- Age of pilot
- Licence Type held
- Pilot's total flying hours
- Pilot's hours on this aircraft type
- Pilot's total flying hours in last 90 days
- Pilot's total flying hours in last 90 days on this aircraft type
- Weight and balance at takeoff
- Terrain where incident occurred
- Weather at time of incident
- Cause of loss
- Injury to passengers/pilot
- Other comments
- Assessor recommendations
- Additional information.

Data were extracted by one researcher and entered directly into a purpose built Microsoft Access database. Narrative data (incident, synopsis, introduction, circumstances) were coded into new categorical fields: purpose of flight; phase of operation (at which the incident occurred); effect on flight; and outcome of incident (up to three outcomes coded).

4.1.4 Analysis Procedure and Resources Invested

The HFACS analysis involved the use of an expert panel of experienced aviators to analyse the cases. The panel comprised aviators with considerable experience in both instructing pilots and investigating accidents and incidents in safety-critical domains, including aviation, and had existing knowledge about system approaches to accident causation. Nonetheless, the panel was subjected to a training programme to ensure that each member's understanding of accident causation and the HFACS method was consistent.

Each case was analysed by at least two of the analysts from the panel. An experienced insurance assessor also analysed all cases using HFACS. Both analyses (panel member and insurance assessor) were then compared for reliability analysis purposes.

4.1.5 Outputs

Further investigation revealed that nineteen of the cases involved storm damage or were not investigated by the insurer. These cases were not analysed using HFACS. Table 4-6 presents the HFACS codes identified as playing a role in the 169 incidents analysed.

Table 4-6   Frequency and percentage of HFACS codes identified in GA incidents

HFACS level	Subcategory	Frequency	% of all incidents
Organisational influences	Organisational process	6	3.6
	Organisational climate	0	0
	Resource managemet	6	3.6
Unsafe supervision	Supervisory violation	6	3.6
	Failed to correct a known problem	5	3
	Planned inappropriate operations	3	1.8
	Inadequate supervision	11	6.5
Preconditions for unsafe acts	Conditions of operators	75	44.4
	Adverse mental states	55	32.5
	Adverse physiological states	3	1.8
	Physical/mental limitations	32	18.9
	Personal factors	29	17.2
	Crew resource management	9	3.3
	Personnel readiness	21	12.4
	Physical environment	37	21.9
	Technological environment	3	1.8
Unsafe acts	Violation	27	16
	Perceptual errors	27	16
	Skill-based errors	103	60.9
	Decision errors	60	35.5

Almost three-quarters of the cases involved one or more unsafe acts by aircrew (69 per cent). Skill-based (61 per cent) and decision errors (36 per cent) were the most common categories of unsafe act, followed by perceptual errors (16 per cent) and violations (16 per cent). A quarter of the cases analysed did not involve any identifiable human error. These cases included mechanical factors, e.g. when a power loss occurred that resulted in a forced landing, yet no human error could be identified, and lightning strikes, again where it could not be determined if the aircrew adhered to weather advisories, and so on.

Preconditions for unsafe acts were found in almost 60 per cent of incidents, predominantly involving condition of operator factors (44 per cent) and environmental factors (24 per cent). Unsafe supervision and organisational factors were found to be present only in a small proportion of the cases analysed.

Only limited failures were found at the unsafe supervision and organisational influences level. The unsafe supervision failures included supervisory violations (in 3.6 per cent of incidents), failure to correct a known problem (3 per cent), planned inappropriate operations (1.8 per cent) and inadequate supervision (6.5 per cent). The organisational influences level failures included organisational process (3.6 per cent) and resource management (3.6 per cent) failures.

The analysis can also be categorised by purpose of flight (Table 4-7). Unsafe acts, particularly skill-based errors, were more frequent during private, training, and agriculture operations. Within the skill-based error category, errors related to 'poor technique/airmanship' were more frequent for training (67 per cent) and agricultural (59 per cent) operations, compared to private (38 per cent) and charter (22 per cent). Examples of the skill-based errors involved included mishandled takeoff; pilot did not reduce power when aircraft deviated from strip; pilot failed to control airspeed during forced landing; and failed to correct undershooting on the approach. Also within the skill-based errors, the error 'failure to see and avoid' was more frequent in private (22 per cent) and agriculture (27 per cent) operations. Examples of this error type included 'taxied into tyre marker when trying to avoid horses on the aerodrome' and 'pilot failed to see and avoid power lines'.

Decision errors were more frequent in training and agricultural operations. In particular, decisions to undertake inappropriate manoeuvres/procedures were prominent in agricultural operations (36 per cent) (e.g. operating close to wires in wind conditions that made wire avoidance difficult, incorrect transition to forward flight resulted in rotor speed decay), and decisions to undertake tasks that exceeded abilities were common in training scenarios (33 per cent) (e.g. the attempted lift-off technique was beyond the student's ability, student pilot not ready for first solo). Perceptual errors of misjudgement were more common in training flights (43 per cent). Violations were found to be more frequent for agriculture and other aerial work.

Preconditions for unsafe acts were significant factors for all flights, but especially for private, training, and agricultural. For example, under condition of operator factors, 'loss of situational awareness' was noted in private (13 per cent), training (10 per cent), and agricultural (14 per cent) operations. In addition, for private flights poor flight vigilance and inadequate experience were identified in 13 per cent and 11 per cent of incidents respectively. For training flights, instances of overconfidence (14 per cent) and inadequate experience for the complexity of the situation (33 per cent) were also identified. For agricultural operations, complacency (14 per cent), distraction (9 per cent), and inadequate experience (9 per cent) were found. Similar to the findings of Wiegmann and Shappell (2003), unsafe supervision and organisational factors were mainly evident in training and scheduled flights, namely those that do not involve owner operators.

Since the aim of the analysis was to validate HFACS as an appropriate approach for analysing GA aviation accident and incidents, it was important to evaluate the reliability of the analysis produced. In this case the signal detection paradigm was used. This has been used in the past for evaluating the reliability of error prediction and classification approaches and provides a simple means of assessing the reliability and sensitivity of error or accident analysis methods (e.g. Harris et al. 2005; Stanton et al. 2009). Here analyst outputs are compared and a sensitivity index score based on hits (error identified by both analyst and expert assessor), false alarms (error identified by analyst but not expert assessor), misses (error identified by expert assessor but not identified by analyst) and correct rejections (error correctly not identified from error taxonomy). The signal detection matrix is presented in Figure 4-1. The formula for calculating sensitivity index scores is presented in Formula 1.

Table 4-7 HFACS analysis by purpose of flight

	Private N	(%)	Charter N	(%)	Aerial N	(%)	Fly training N	(%)	Agricultural N	(%)	Scheduled flight N	(%)	Other/missing N	(%)	Total N	(%)
UNSAFE ACTS	50	78	10	45	12	60	19	90	19	86	5	36	2	33	117	69
Errors	48	75	10	45	12	60	19	90	18	82	5	36	5	83	114	67
Skill-based errors	41	64	10	45	9	45	18	86	18	82	5	36	2	33	103	61
Decision errors	23	36	5	23	7	35	10	48	12	55	2	14	1	17	60	36
Perceptual errors	9	14	1	5	3	15	9	43	2	9	2	14	1	17	141	83
Violations	10	16	2	9	5	25	1	5	8	36	0	0	1	17	27	16
PRE-CONDITIONS FOR UNSAFE ACTS	43	67	9	41	9	45	15	71	15	68	8	57	1	17	100	59
Conditions of operators	34	53	5	23	6	30	13	62	11	50	5	36	1	17	75	44
Personnel factors	15	23	0	0	2	10	8	38	1	5	2	14	1	17	29	17
Environmental factors	16	25	5	23	4	20	3	14	5	23	6	43	1	17	40	24
UNSAFE SUPERVISION	4	6	2	9	1	5	6	29	4	18	4	29	0	0	21	12
Inadequate supervision	2	3	0	0	1	5	5	24	2	9	1	7	0	0	11	7
Planned inappropriate operations	1	1	0	0	0	0	1	5	0	0	1	7	0	0	3	2
Failed to correct a known problem	1	1	0	0	0	0	0	0	1	5	3	21	0	0	5	3
Supervisory violation	1	1	2	9	0	0	0	0	2	9	1	7	0	0	6	4
ORGANISATIONAL INFLUENCES	2	3	2	9	2	10	2	10	2	9	2	14	0	0	12	7
Resource management	1	1	2	9	0	0	2	10	1	5	0	0	0	0	6	4
Organisational process	1	1	0	0	2	10	0	0	1	5	2	14	0	0	6	4
Additional factors (no error in four levels of HFACS)																
Mechanical factors	10	15	7	32	5	25	2	10	3	14	1	7	2	33	30	18
Maintenance factors	6	9	0	0	4	20	2	10	3	14	1	7	1	17	19	11
TOTAL	64		22		20		21		22		14		6		169	

		Analyst classification	
		Yes	No
Expert assessor	Present	HIT	MISS
	Absent	FALSE ALARM	CORRECT REJECTION

Figure 4-1   Signal detection paradigm matrix

$$Si = \frac{\left(\frac{Hit}{Hit + Miss}\right) + 1 - \left(\frac{False\ Alarm}{FA + Correct\ Rejection}\right)}{2}$$

Formula 1   Sensitivity index formula

In this case each output provided by the two analysts who analysed each case were compared to the experienced assessors' analysis outputs and a general agreement score and sensitivity index score was calculated. The results are presented in Table 4-8. At the unsafe acts and preconditions for unsafe acts levels, the level of general inter-rater agreement was found to be over 80 per cent for both comparisons. Again at both levels, sensitivity index values of between 0.72 and 0.79 were found, which represents a high level of reliability for the HFACS analyses offered by the different analysts.

Table 4-8   Agreement in error classifications between assessors

	Unsafe Acts		Preconditions for unsafe act	
Assessor Pairings	% Agreement	SI	% Agreement	SI
Pair 1	83.9	0.79	88.3	0.73
Pair 2	86.1	0.76	88.8	0.72

4.1.6 Discussion

The analysis presented gave an insight into the nature of GA accidents and incidents and also into the difficulties associated with the use of insurance data for accident analysis purposes. Regarding the cases analysed, skill-based errors were the most commonly identified unsafe act involved, being present in over 60 per cent of the cases analysed, followed by decision errors, which were present in over 35 per cent of all cases. At the preconditions for unsafe acts level, various preconditions were identified, with the condition of operators (44.4 per cent), physical environment (21.9 per cent), physical and mental limitations (18.9 per cent) and personnel factors (17.2 per cent) being the most popular. Unfortunately, few factors at the inadequate supervision and organisational influences levels were identified; however, there are various reasons for this. For example, it is likely that since GA incidents often involve owner operators, there is little scope for supervisory or organisational failures (Wiegmann and Shappell 2003). Further, this is also likely to be a function of the data available which are shaped by the focus of insurance investigations.

The findings derived from this case study were compared with other aviation-based HFACS analyses presented in the academic literature (see Table 4-9).

The comparison presented in Table 4-9 shows that the findings are broadly consistent with the findings from the other studies considered. Notably a slightly lower proportion of unsafe acts were observed in this study compared to Wiegmann and Shappell (2003); however, there are several potential reasons for this. The most obvious is that unlike the US studies the present analysis included cases that did not include aircrew error, which means that the proportion of unsafe acts reported here will be lower than noted for other studies. Also relevant to note here is that the quality of Human Factors data in existing insurance files is inferior to that of NTSB (National Transportation Safety Board) reports, which are the source of data for most US HFACS studies.

In addition to the positive inter-rater reliability findings reported, the comprehensiveness of the HFACS method was demonstrated in that few failures were classified under 'other' due to there being no appropriate HFACS factor. That said, two themes that did not appear to be catered for by the HFACS method were currency (there were several reports in which the pilot's currency, as distinct from expertise, was noted) and an inadequate landing/parking surface. Given that HFACS captured almost all of the failures identified by the experts, it is recommended that any modifications to the HFACS framework should not be made until a much larger sample of data is analysed.

While only a very small number of cases were listed as 'insufficient data to analyse', it must be noted that the data collected by aviation insurers do lack the depth and focus to facilitate optimal HFACS analysis. While the analysis using HFACS presented here yielded valuable data, and in fact the type of data not previously published for insurance claims in Australia, there is little doubt that improving the quality of the data collected would significantly improve the agreement between raters when using analysis methods such as HFACS (cf. Wiegmann and Shappell 2003).

Table 4-9   Comparison of case study findings with selected HFACS aviation analyses

	Wiegmann and Shappell (2001)	Gaur (2005)	Shappell and Wiegmann (2001)	Shappell and Wiegmann (2003)	Wiegmann and Shappell (2003)		Present analysis
	Commercial	Commercial	GA - CFIT	GA accidents	GA Fatal	Non-F	GA incidents
UNSAFE ACTS		77					69
Errors							
Skill-based errors	60	52	49	74	82	≈80	61
Decision errors	29	22	45	35	36	≈40	36
Perceptual errors		15	31	8	12	≈10	16
Violations	27			14	32	10	16
PRE-CONDITIONS FOR UNSAFE ACTS		48					59
Conditions of operators		44					44
Adverse mental states	13	13		5			33
Adverse physiological states	2	4	13	3			2
Physical/mental limitations	11	31		18			19
Personnel factors		19					17
Crew resource management	29	13		11			5
Personnel readiness		8		2			12
Environmental factors							24
Physical environment							22
Technological environment							2
UNSAFE SUPERVISION	16[†]	25	Very few	Not noted			12
Inadequate supervision		15					7
Planned inappropriate operations		8					2
Failed to correct a known problem		4					3
Supervisory violation		4					4
ORGANISATIONAL INFLUENCES		52	Very few	Not noted			7
Resource management		40					4
Organisational process		42					4

[†] This represents the number of cases involving unsafe supervision or organisation influences

4.1.7 Summary of Key Findings

The case study presented involved the application of HFACS for the analysis of Australian GA insurer accident and incident data. In summary, the analysis demonstrated that, for the cases analysed:

- Skill-based errors were the most commonly involved unsafe act, being present in over 60 per cent of the cases analysed. Decision errors were the next most common unsafe act, present in over 35 per cent of all cases. This finding is broadly consistent with the findings derived from other HFACS civil and military aviation analyses (e.g. Gaur 2005; Shappell and Wiegmann 2001; Wiegmann and Shappell 2001);
- Various preconditions for unsafe acts were identified in the cases analysed, with the condition of operators (44.4 per cent), physical environment (21.9 per cent), physical and mental limitations (18.9 per cent) and personnel factors (17.2 per cent) being the most popular;
- Few factors were identified at the inadequate supervision and organisational influences levels, with the lack of supervisory systems in place in GA and the data available being possible explanations for this;
- Analysis of inter-rater reliability revealed acceptable levels of agreement between the analysts involved.

4.1.8 Acknowledgements

The overall research programme under which the case study presented was undertaken was sponsored by the Aviation Safety Foundation Australasia and supported by the Australian Transport Safety Bureau and BHP Billiton. The authors would like to acknowledge the assistance received from the Project Steering Committee, represented by the following organisations: the Australian Transport Safety Bureau; QBE Aviation; Vero Aviation; Asset Insure; and BHP Billiton. The authors would also like to thank Gary Lawson-Smith and Russell Kelly JP from the Aviation Safety Foundation Australasia for their support in undertaking this project and also Richard Gower, Clive Philips, and Geoff Dell for their invaluable contributions through their involvement in the expert panel. Finally, the authors would like to acknowledge Dr Michael Fitzharris, Karen Ashby, Carolyn Staines, and Ashley Verdoorn for their involvement in the research undertaken.

4.2 HFACS ANALYSIS OF MINING INCIDENTS

4.2.1 Introduction

Epidemiological studies have shown that mining workers face a relatively hazardous work environment compared to workers in other industries. For example, based on a comparison of work-related fatal injuries occurring in various industries in the US, Australia and New Zealand during the 1980s and 90s, the rate of fatal injury for mining workers was found to be between seven and ten times that of the average worker in the population (Feyer et al. 2001). As well as the pain and trauma associated with injury, the financial cost of medical care and lost productivity due to injury in mining is also considerable. For example, in an analysis of US data from 1993, 'lignite and bituminous coal mining' was ranked second in terms of average cost per worker for fatal and nonfatal injuries (Leigh et al. 2004).

Previous studies have identified factors that appear to have some relationship to injury severity and risk in mining workers. These include worker experience, the equipment used, and the environment in which mining activities are undertaken. Non-fatal injuries most often involve non-powered hand tools, while fatalities most often occur during off-road ore haulage activities (Groves et al. 2007). Trucks, conveyors, and front-end-loaders have been found to account together for 40 per cent of fatalities (Kecojevic et al. 2007). Overhead power lines have also been found to be a major causal factor in fatal electrical accidents (Cawley 2003), and underground mining exhibits a higher rate of injuries and fatalities than surface operations (Karra 2005). The role of worker experience has also been investigated with mixed outcomes. While some studies report that most injured workers have less than five and ten years' experience respectively (Groves et al. 2007; Kowalski-Trakofler and Barrett 2007), others claim a similar level of risk for inexperienced and experienced workers (Bennett and Passmore 1985; Maiti and Bhattacherjee 1999). For example, experienced electrical workers have been found to deliberately engage in risky behaviours (e.g. a short-cut), while knowing that they were potentially dangerous (Kowalski-Trakofler and Barrett 2007). Finally, workers' perceptions of variables including job dissatisfaction, poor management commitment, time pressures, and concerns with policies, have also been found to be important predictors of work injury and the propensity to commit violations (Paul and Maiti 2008).

Taken together the literature indicates that a range of factors, related to the individual and the wider organisational system, play a role in mining accidents. The development of effective mining accident countermeasures must therefore be based on an understanding of both the individual worker factors and higher level organisational and managerial factors involved in accidents. As part of their ongoing efforts to increase safety through more informed countermeasure development, a major Australian mining company involved in both surface and underground mining approached the Monash University Accident Research Centre to undertake a systems-based analysis of their 'significant incident' data for the financial year 2007/8. Accordingly, the HFACS approach was used to analyse 267 significant incidents from this period. The aim of the analysis was to identify both the individual and system-wide causal factors involved in the incidents and to examine the associations between failures across the organisational levels of the mining system, with a view to generating appropriate incident prevention and countermeasure strategies.

4.2.2 Incident Description

Data were provided by the mining company on all significant incidents for the financial year 2007/8. The dataset was restricted to those incidents that were classified as a potential Level 4 ('Major') or potential Level 5 ('Critical') incidents. A major incident is defined as: a fatality, permanent disabling injury, or damage greater than $1M. A critical consequence is defined as: multiple fatalities, or damage greater than $10M. In total, there were 267 such incidents.

CHAPTER 4 • THE HUMAN FACTORS ANALYSIS AND CLASSIFICATION SYSTEM

4.2.3 Data Sources and Data Collection

The research team was provided with the individual case files that contained incident narratives, photographs, and accident analysis outputs derived from the mining companies' existing form of analysis, known as the Incident Cause Assessment Method (ICAM; BHP Billiton 2005). The research team also visited one of the company's major mining sites to speak with SMEs and observe both surface and underground mining operations.

4.2.4 Analysis Procedure and Resources Invested

Data storage A database was developed to store the de-identified data extracted from the accident reports. Descriptive accident data were extracted, such as commonly reported variables including, but not limited to, the event type, site involved, repeat events, the injury mechanism and the injury severity.

Data coding Since the data was provided to the research team in the form of ICAM analysis outputs, a two-stage coding procedure was required. First, to ensure the validity of the original ICAM analyses, the data had to be re-coded by ourselves against the ICAM framework. Second, for the HFACS analysis, the revised ICAM codes then had to be mapped onto the HFACS framework. In accordance with accepted scientific practice to maximise data reliability (e.g. Li and Harris 2006), a panel of experienced Human Factors professionals was assembled to analyse the data. These experts had vast experience in the use of error frameworks to code and analyse safety data, and had existing knowledge about systemic approaches to human error and accident causation. In total, there were 2,868 ICAM codes listed across the 267 cases. The levels of agreement in the coding of the ICAM data were as follows: individual/team actions ($k = 0.71$); environmental ($k = 0.80$); environmental ($k = 0.81$); and organisational ($k = 0.79$). Subsequent to the independent checking, the raters met to discuss any discrepancies between their judgements, with any discrepancies being resolved through further discussion and mutual agreement. Having examined all of the ICAM codes, and resolved the assignment of ICAM codes to incident descriptions, the second task was to ascertain how the ICAM framework could be mapped onto the HFACS framework (Wiegmann and Shappell 2003). Two Human Factors specialists independently mapped each of the ICAM factors to one of 17 HFACS categories. The few disagreements identified were resolved through discussion with a third Human Factors expert.

Statistical analysis Preliminary assessments of the incident characteristics and HFACS data were performed using frequency counts. The nature of the relations, if any, between each HFACS level with the level immediately above was examined using Fisher's Exact Test for contingency tables. Odds ratios (OR) were calculated to assess the strength of association. Odds are calculated for lower level factors – the odds is the ratio of the probability that a (lower-level) factor is present, to the probability that it is absent. The odds were calculated under two conditions: one for when a higher-level factor is present, and another for when a higher-level factor is absent. An odds ratio was then calculated by dividing these two odds. Two standards of statistical significance were set: $p < .05$, and p < .005. Analyses were conducted using the software package R (R Development Core Team 2007).

4.2.5 Outputs

Incident classification   The final sample for analysis comprised 263 cases. Figure 4-2 shows a breakdown of incidents by event type.

Figure 4-3 shows a breakdown of the incidents by activity, describing the type of mining activity involved. The most common activity within the incident data set was Surface Mobile Equipment (38 per cent), followed by Working at Heights (21 per cent).

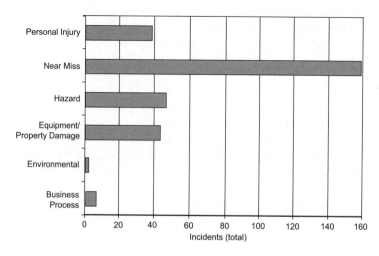

Figure 4-2   **Number of significant incidents by event type**

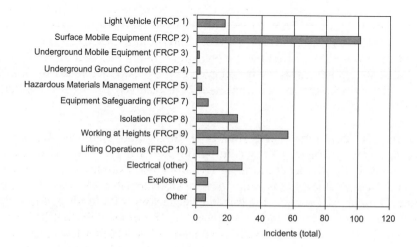

Figure 4-3   **Number of significant incidents by mining activity**

CHAPTER 4 • THE HUMAN FACTORS ANALYSIS AND CLASSIFICATION SYSTEM

HFACS analysis outputs Table 4-10 presents the HFACS codes identified in the 263 incidents analysed. Ninety per cent of incidents involved one or more unsafe acts. The most frequent category of unsafe act was skill-based error (64 per cent), followed by violation (57 per cent). The skill-based errors identified commonly involved failure to identify the hazards involved in completing a task or the incorrect use of equipment, while the violations identified typically involved a failure to follow organisational procedures such as failing to use personal protective equipment.

One or more preconditions for unsafe acts were identified in almost 90 per cent of incidents. The preconditions most commonly involved were the physical environment (56 per cent), typically involving poorly accessible work areas and/or abnormal weather, lighting and other environmental conditions, followed by the technological environment (33 per cent), commonly unsound or poorly designed equipment. Adverse mental states and mental or physical limitations were also prevalent preconditions (both 25 per cent), typically involving employees being complacent, inexperienced or distracted from the task at hand.

Table 4-10 Frequency and percentage of HFACS codes identified in mining incidents

HFACS level	Subcategory	Frequency	% of all incidents
Organisational influences	Organisational process	172	65.4
	Organisational climate	75	28.5
	Resource management	77	29.3
Unsafe supervision	Supervisory violation	11	4.2
	Failed to correct a known problem	0	0
	Planned inadequate operations	87	33.1
	Inadequate supervision	59	22.4
Preconditions for unsafe acts	Technological environment	86	32.7
	Physical environment	147	55.9
	Crew resource management	40	15.2
	Physical/mental limitations	67	25.5
	Adverse physiological states/personal readiness	28	10.6
	Adverse mental states	66	25.1
Unsafe acts	Violations	151	57.4
	Perceptual errors	0	0
	Skill-based errors	168	63.9
	Decision errors	89	33.8

Unsafe supervision factors were present in 44 per cent of incidents, the most common factor being planned inadequate operations (33 per cent) which commonly involved lack of communication or co-ordination between and within groups, poor preparation and time pressure.

Over 80 per cent of incidents involved organisational factors with organisational process (65 per cent) being the most commonly identified factor at this level. Organisational process issues tended to involve lack of, or substandard formal processes, such as lack of effective Job Safety Analysis (JSA) tasks or lack of formal incident reporting processes.

The statistically significant odds ratios (ORs) and 95 per cent confidence intervals (CIs) for higher level factors (first column) associated with factors at lower levels (second column) are presented in Table 4-11. The 95 per cent confidence interval gives an estimate of the plausible range of values within which the true odds ratio lies, and an odds ratio of one implies no association. We also assessed confidence in the presence of an association by the p-value: one asterisk represents significance at the conventional .05 level, whereas a double asterisk represents significance at the .005 level, which indicates very high confidence that the association is real.

Figure 4-4 presents these associations in graphical form. It is important to note that the absence of an association does not necessarily imply that a particular HFACS category is unimportant. For example, issues with organisation process occurred in over half of the incidents, and it would be desirable to minimise the frequency of such influences. Rather, the significant associations suggest that, given finite resources, more attention should be directed towards those factors that are more likely to lead to eventual unsafe acts or operations.

Table 4-11   Odds ratios for the associations between codes from each HFACS level

HFACS level		OR*/**	95%CI
Preconditions	**Unsafe acts**		
Adverse mental states	Decision errors	1.94*	1.05–3.59
Adverse mental states	Violation	2.01*	1.07–3.86
Adverse physiological states	Skill-based errors	3.78*	1.24–15.4
Crew resource management	Violation	2.17*	1.00–5.08
Physical environment	Violations	0.51*	0.30–0.87
Technological environment	Decision errors	2.84**	1.60–5.06
Unsafe supervision	**Preconditions**		
Inadequate supervision	Adverse mental states	2.15*	1.09–4.2
Inadequate supervision	Crew resource management	4.11**	1.9–8.89
Planned inappropriate operations	Crew resource management	2.96**	1.41–6.28
Supervisory violation	Adverse physiological states/ personal readiness	5.37**	1.07–22.3
Organisational influences	**Unsafe supervision**		
Resource management	Inadequate supervision	1.95*	1.01–3.37
Organisational climate	Inadequate supervision	3.3**	1.72–6.36

CHAPTER 4 • THE HUMAN FACTORS ANALYSIS AND CLASSIFICATION SYSTEM

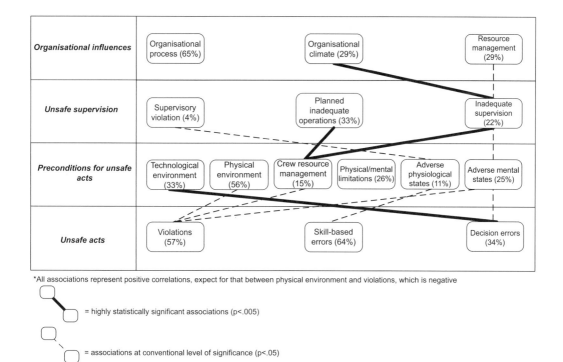

Figure 4-4 Associations between HFACS codes at each level

The analysis of associations highlights key links between failures at the higher organisational levels and the errors made by workers at the sharp end. For example, the results imply that the odds of making a decision error in an adverse mental state are about twice as high as when in a normal mental state. Similarly, adverse mental states and poor crew resource management is associated with more frequent violations, while adverse physiological states is associated with more skill-based errors.

Poor technological environment predicted more decision errors, and this effect is both strong (OR = 2.84) and highly statistically significant (p<.005). We illustrate with two examples. In one incident, the technological environment factor represented: 'Table guards are pull on/off not hinged', and 'Lugs have been welded on handles of the rod slips (previously)', which was associated with the decision error: 'Drillers assistant reached in to where the drive was rotating, once the rod lifted to grab slip (automatic reflex / reaction)'. In a second incident, the technological environment factor was 'suitability of truck mounted of box design is not suitable for the task', which was associated with the decision error: 'The rigger has adapted a work method to protect his lower back which involves positioning himself on top of the truck mounted chain box'. In these and other cases, unsound or poorly designed equipment resulted in decisions to use work methods to complete tasks other than the methods workers would normally adopt.

Several factors at the level of unsafe supervision were associated with various preconditions for unsafe acts. For example, inadequate supervision predicted more adverse mental states and poor crew resource management, and supervisory violations predict more adverse physiological states. Poor crew resource management was predicted by inadequate supervision and planned inappropriate operations. Both of these associations were strong (OR = 4.11 and 2.96) and highly significant ($p < .005$). In one incident, inadequate supervision was described by: 'There was no management or control of the tray coming onto site', which was associated with poor crew resource management: 'There was no communication between the contract transport company and site contacts'. In general, the analysis indicates that when management and planning of personnel and work operations is inadequate, communications and teamwork on site tends to suffer.

Two organisational influences (resource management and organisational climate) were associated with inadequate supervision. In particular, organisational climate was a strong predictor (OR = 3.3) and highly significant ($p < .005$).

4.2.6 Discussion

The following HFACS categories were prevalent in the 263 cases analysed: skill-based errors and violations, physical environment, and organisational processes. To enable a deeper level of analysis, the associations between HFACS categories at the different levels were analysed statistically in order to identify which higher-level factors predicted specific lower-level factors. Through this analysis several significant associations were found, with the following four representing very strong associations: (1) technological environment and decision errors; (2) planned inappropriate operations and crew resource management; (3) inadequate supervision and crew resource management; and (4) organisational climate and inadequate supervision. It was concluded that these associations require greater attention in future accident countermeasure and prevention strategy development, although the other significant associations identified should not be ignored. Also noteworthy is the fact that previous HFACS analyses undertaken in other domains (e.g. aviation, healthcare) have identified similar associations. For example, in an analysis of civil aviation incidents, Li and Harris (2006) found associations between organisational climate and inadequate supervision, as well as between planned inappropriate operations and resource management. Also in the civil aviation domain, Li et al. (2008) found that inadequate supervision predicted crew resource management, which predicted violations. In general aviation, Lenné et al. (2008) found that both crew resource management and adverse mental states predicted violations, as was found in the present study. Finally, in the context of cardiovascular surgery, El Bardissi et al. (2007) found associations between planned inappropriate operations and crew resource management, as well as organisational climate and inadequate supervision. The fact that these associations appear to generalise across industries suggest that such relationships may reflect common properties of safety critical systems and their effects on human error at the sharp end.

Skill-based errors One of the more common categories of error found during this analysis was skill-based errors. This is consistent with the findings derived from HFACS analyses undertaken in other domains, including aviation (Wiegmann and Shappell 2003) and rail (Baysari et al. 2008). Skill-based errors arise during skill-based behaviour, which is characterised by highly practised and automatic behaviour where there is only small conscious control on behalf of the operator (Vicente 1999). Skill-based errors typically include failures of attention and memory, technique errors and omissions (Wiegmann and Shappell 2003). Typical countermeasures for skill-based errors include training, warning systems and redesign of equipment to improve the system's ability to absorb the consequences of human error (Shappell and Wiegmann 1997). Reason (2002) also suggests that the provision of timely and suitably located reminders will lead to reductions in omissions (e.g. omitting a step in a procedure), and outlines ten criteria for effective reminders. Reason (2002) also discusses the use of forcing functions, whereby mechanical or electronic devices that block onward actions until all required steps have been completed, are added to systems.

Of course, systems theory advocates such as Reason (1990) and Wiegmann and Shappell (2003) advocate the removal of latent failures as the most appropriate approach for dealing with errors. Focusing more on the overall system then, a statistically significant association was found between adverse physiological states and skill-based errors, which indicates that factors such as fatigue are leading to skill-based errors being made. This supports previous findings in the Human Factors literature which show an association between skill-based errors and adverse physiological states (generally, fatigue). Furthermore, adverse physiological states are, in the present data, associated with supervisory violations. This suggests that skill-based errors may be reduced if fatigue-related issues are addressed at the supervisory level. This might include providing supervisors with training in strategies for detecting and managing fatigue. It is notable, e.g. that the lack of a procedure for dealing with worker fatigue once it is identified was reported in the data. In addition to countermeasures at the supervisory level, employees also have a reciprocal duty of care regarding fatigue management, and steps that they can take to better manage fatigue include developing self-awareness of known patterns of fatigue, taking frequent breaks, engaging in social interaction, communication and other activities, and varying their schedules (Stanton, Salmon et al. 2010). It is notable that most of these steps are likely to be achieved only through support from supervisors.

Violations The frequency of violations found in this sample is relatively higher than that found through HFACS analyses in other industries such as aviation (e.g. Lenné et al. 2008). Various forms of violation are discussed in the literature. Deliberate violations are characterised by operators deliberately deviating from set rules or procedures, whereas unintentional violations refer to those instances when an operator unintentionally deviates from a set of rules or procedures. Unintentional violations can be treated through increasing awareness of violations via supervisor input, workshops, educational programmes and signage informing workers of those activities that constitute violations. Deliberate violations, however, are more difficult to address. Reason (1997) distinguishes between three types of deliberate violation: routine, optimising, and necessary violations. Routine violations involve taking short-cuts through procedures in order to achieve a particular task. Optimising violations involve the optimisation of non-functional goals during task performance. Reason (1997) cites the example of a driver (wishing to get from A to B) who optimises speed and indulges in aggressive instincts during the journey from A to B. Necessary violations involve essential deviation from the rules in order to achieve a particular task. For example, when a routine procedure in the workplace is not working, the use of a non-routine procedure to get the job done would be termed a necessary violation. The problem with deliberate violations is often that they have become accepted components of procedures, and even supervisors and management often tolerate them as they often get the job done. They may even be passed on to new workers through on-the-job training which leads to a lack of awareness that they are in fact violations and not the norm (Wenner and Drury 2000). The important thing to note about violations is that they take perpetrators 'closer to the edge', increasing the chance that subsequent errors will have damaging outcomes (Reason 1995).

Most deliberate violations can be eliminated by ensuring that the procedures represented are the easiest and most efficient (Wenner and Drury 2000). This may involve an evaluation and redesign of the procedures that are prone to violations. Violations can also be prevented by ensuring that appropriately designed and well maintained equipment is provided to workers. In addition, increasing awareness of what does and does not constitute violation activity is useful.

Supervision Inadequate supervision was found to be present in 44 per cent of the incidents analysed. Recurrent issues that emerged from the data were the provision of inadequate instructions by supervisors, a lack of on-the-job supervision, inadequate supervision of tasks, poor hazard identification on the part of supervisors, and acceptance of inappropriate practices (i.e. violations). The efficiency of an organisation's supervisory arrangements is a key factor in system safety and performance, with various studies across the safety critical domains previously identifying poor or inadequate supervision as a key contributory factor in accidents and incidents (e.g. Bomel Consortium 2003; Brazier et al. 2004 etc). The high presence of unsafe supervision factors in the incidents analysed indicates that further investigation is required in order to assess the supervisory arrangements currently in place at the mines analysed. The importance of organisations' monitoring and assessing their supervisory approach is well versed in the literature, and our findings suggest that some sort of assessment of the organisations' current supervisory arrangements is required. Ward et al. (2004), e.g. argue that, organisations need to understand their own approach to supervision and its weaknesses so that they can take effective countermeasures. Brazier and Ward (2004) present a supervision assessment methodology that enables organisations to evaluate their supervisory arrangements and the subsequent impact on health and safety.

Based on a review of the literature, Stanton, Salmon et al. (2010) proposed a series of guidelines for enhancing supervision within the safety critical domains. These included: clearly defining the competencies required for each supervisory role; ensuring the appropriate individuals are selected for supervisory roles; providing appropriate training based on the competencies required; clearly defining and communicating the roles and responsibilities of each supervisory role to supervisors, subordinates, and managers; continual monitoring and assessment of the supervisory system in place; encouragement of communication and high levels of interaction between supervisors, subordinates and management; clearly defined and expressed lines of reporting; optimisation of supervisor workload; and the avoidance of demarcation. From these guidelines, clearly defining and communicating the role of supervisors throughout the organisation, assessing the supervisory system and encouragement of clear communication and high levels of interaction between supervisors and subordinates appear to be particularly pertinent in this case.

Organisational processes Organisational process failures were found in 65 per cent of the incidents analysed. Prominent failures of organisational processes were related to procedures (e.g. inadequate procedures, no procedures available, or poor awareness of procedures), hazard identification and risk assessment (e.g. failure to identify hazards, hazard identification/risk assessment not undertaken, and a lack of an appropriate risk assessment procedure/tool), and inadequate, or lack of, work instructions. A significant portion of the failures at the organisational process level can therefore be treated by focusing on the procedures currently used.

Procedural failures have previously been identified as problematic. Marsden (1996), e.g. states that, in the nuclear power domain, procedural failures are involved in around 70 per cent of accidents and incidents (e.g. INPO 1986, Goodman and DiPalo 1991, cited in Marsden 1996). Various forms of procedural failure have been identified, including (adapted from Green and Livingston 1992, cited in Marsden 1996):

1. Technical accuracy or completeness of the procedure at fault;
2. Poor formatting of the procedural document;
3. Language or syntax problems associated with the procedural document;
4. Inadequate procedure development process;
5. Procedure is poorly located or cross referenced;
6. Insufficient verification or validation of the procedural document;
7. Failure to revise the procedure;
8. Training and procedures poorly interrelated.

Marsden (1996) views procedural failures from an organisational perspective, suggesting that the factors involved are more likely to be found within the infrastructure of the organisation as opposed to the fallibility of the human operators involved. Marsden (1996) suggests that there are three basic failure types involved in procedural failures. These include weaknesses in the procedure preparation process, problems emerging from the failure of employees to follow procedures, and breakdowns within the infrastructure of the procedure system.

One of the main problems associated with procedures is operators' perceptions of them. According to Cox and Cox (1996) procedures are often perceived by workers to be inaccessible, ambiguous, overly demanding on memory and inadequate when used to diagnose plant problems; further Ockerman and Pritchett (2001) point out that procedures are often seen to be inefficient, burdensome, overly difficult or plain wrong. They also suggest that procedures may often be intentionally ignored since workers perceive more drawbacks in following the procedure than benefits.

Developing procedures is a complex process involving many activities including identifying the tasks that require procedural support, determining the level of assistance required, determining the type or format of procedure required, writing, reviewing and reiterating the procedure, and then obtaining approval for the procedure. Following official publication of the procedure, staff should receive appropriate training in the procedure, and the procedure should be regularly reviewed and updated when required. A formal procedural development process is thus required; however, Marsden (1996) suggests that procedures are often developed using informal methods that typically rely on the knowledge of individuals familiar with the system but not with the actual behaviours required to operate it. The end results are often incomplete, incorrect or generally unrealistic procedures. The HSE's guide for reducing error and influencing behaviour (HSE 1999) does give some guidance, suggesting that procedures should contain the following information: purpose, precautions that must be observed, tools and equipment required, conditions to be satisfied before commencing, documentation needed (e.g. manuals) and a description of the procedural steps required. The HSE (2008) also outlines various key principles of procedure design, including using a format, style and level of detail that is appropriate to the task, user and consequences of failure, using task analysis methods for procedural content identification, encouraging compliance through involvement of end-users, designing out procedural violations and considering the links between procedures and competencies.

Assessing the procedures currently in place is also critical. It is vital that both the procedures themselves and also the processes that the organisation has in place with regard to procedure development, writing, approval, training and maintenance, are assessed. For the procedures themselves, Ockerman and Pritchett (2001) suggest that the efficacy of a procedure can be determined by measuring the following attributes:

1. Comprehensiveness of the procedure. A procedure's comprehensiveness relates to the range of situations, environments and operations that the procedure is applicable to.
2. Detail of the procedure. The level of detail of the procedure relates to the degree to which each step within the procedure specifies the exact actions required by the user.
3. Accuracy of the procedure. The accuracy of the procedure relates to the extent to which completion of the steps specified will produce the desired results.

Recommendations As the overarching aim of this analysis was to suggest potential initiatives to substantially reduce the risk of injury to mining workers, a summary of the countermeasures and strategies proposed on the basis of the HFACS analysis is presented in Table 4-12.

4.2.7 Summary of Key Findings

In conclusion, the analysis presented identified a number of significant themes within the significant incident data for the financial year 2007/8. The analysis indicated that several HFACS categories appeared frequently in the incidents. These included skill-based errors and violations, physical environment, and organisational processes. Based on analysis of the associations between the categories at different levels, the following four strong associations were found: (1) technological environment and decision errors; (2) planned inappropriate operations and crew resource management; (3) inadequate supervision and crew resource management; and (4) organisational climate and inadequate supervision.

In response to the findings derived from our analysis, a series of potential countermeasures, based on the wider Human Factors and safety literature, were discussed. It is notable that more targeted countermeasures could only be proposed on the basis of more exhaustive analyses of the incident data, and also based on the analysis of more incidents. In particular, the analysis of a larger data set containing more incidents (e.g. two or more years' worth of data) is recommended.

4.2.8 Acknowledgement

The authors would like to thank the mining company involved for funding this research. We would also like to acknowledge Dr Charles Liu and Margaret Trotter for their contribution to the research undertaken.

Table 4-12 Summary of recommendations for countermeasures/incident prevention strategies proposed on basis of HFACS analysis

Findings/ Failures	Linked HFACS failures	Proposed countermeasures
Skill-based errors	Adverse physiological states	Development of procedures for detecting and managing fatigue; Supervisor training in procedures for detecting and managing fatigue; Fatigue management workshops; In-depth analysis of skill-based error data.
Violations	Adverse mental states; Crew resource management; Physical environment.	In-depth analysis of violations data; Evaluation/redesign of procedures prone to violations; Evaluation/redesign of equipment prone to violations; Enhance awareness of violation activities.
Decision errors	Adverse mental states; Technological environment.	In-depth analysis of decision error data.
Inadequate supervision	Adverse mental states; Crew resource management.	Assessment of current supervisory system; Redesign of supervisory system; Clear definition and communication of supervisor roles and responsibilities; Encouragement of communication and high levels of interaction between supervisors and subordinates.
Resource management	Inadequate supervision	See inadequate supervision section.
Organisational climate	Inadequate supervision	Assessment of procedures and procedure development and implementation system; Redevelopment of procedure development and implementation system; Redevelopment of selected problem procedures.
High incidence of near misses		In-depth analysis of near miss data.
Invalid ICAM codes	N/A	Redevelopment of ICAMs training procedure; Retraining in ICAMs methodology and coding procedure; Redevelopment of ICAMs methodology and supplementary details section; Human Factors/systems error theory workshops.

5
The Critical Decision Method: Retail Store Worker Injury Incident Case Study

5.1 INTRODUCTION

Warehouse superstore is the term given to a currently expanding branch of retail stores in which the majority of merchandise is stored in the store sales area. Due to the wide range of stock on offer and high requirement for manual handling activities, this domain has been previously identified as potentially hazardous for its workers (St-Vincent et al. 2005). Despite this, the characteristics of activities in these stores are not well known (St-Vincent et al. 2005). Although data exist on the types of injury incidents and their outcomes in terms of injury nature and severity, little is currently known regarding the causal factors involved in injury-causing incidents. One aspect that remains largely unexplored is the factors influencing warehouse superstore workers' decision-making and behaviour, in particular, why store workers decide to violate procedures or rules, resulting in risky behaviours that are known to lead to injury.

The area of violations on the whole is important for accident causation studies. Violations, whereby human workers either intentionally or unintentionally deviate from a set of rules or procedures, have a history of involvement in accidents in the safety-critical domains, with the most famous example being the deliberate violations implicated in the Chernobyl disaster (Reason 1987, cited in Lawton 1998). Despite much being known about different forms of violations (e.g. Reason 1997) and the individual factors which increase the likelihood of them (e.g. Reason 2008), comparatively little research has focused in detail on the causes of violations that reside outside of the individuals involved (Alper and Karsh 2009). This is particularly true of the wider organisational or system factors involved.

Research has focused on the frequency and nature of violations in areas such as driving (e.g. Kontogiannis et al. 2002; Nallet et al. 2010; Shi et al. 2010), motorcycling (Cheng and Ng 2010), aviation maintenance (e.g. Hobbs and Williamson 2002) and healthcare (Patterson et al. 2006), but many other domains, including retail, remain largely unexplored. The aim of this case study was to investigate the factors influencing workers' decisions to engage in violational behaviours known to lead to injury-causing incidents, and to look across to wider systemic factors influencing behaviour in this context.

5.2 INCIDENT DESCRIPTION

Work Context

This study involved a major retail chain in which store workers are required to regularly engage in manual handling activities. The study focused on those workers who typically worked on the shop floor during store opening hours, and had a range of activities to undertake as part of their normal day-to-day job. These include dealing with and assisting customers (e.g. lifting heavy products onto customer trolleys, providing advice regarding different products), processing stock, maintaining cleanliness and tidiness of aisles, and merchandising stock (i.e. facing up). Processing of stock (i.e. moving stock from delivery warehouse to shop floor) via manual handling activities represented the core activity of the workers focused on during the study.

5.3 DATA SOURCES AND DATA COLLECTION

A programme of research incorporating observational study, task and cognitive task analysis, and safety culture assessment was undertaken (only the cognitive task analysis component is reported here). The CDM component involved one researcher visiting 15 different stores and interviewing store workers who had previously been involved in an injury-causing incident as reported in the company 2008/9 accident and injury data. A total of 49 CDM interviews were conducted across the stores visited. Participants were recruited on a voluntary basis, with those store workers working on the day of the observational study who had previously been involved in an injury-causing incident asked if they would like to take part in an interview for a safety research project. For the CDM interviews a set of appropriate interview probes was adapted from the literature on previous CDM applications (e.g. Crandall et al. 2006; O'Hare et al. 2000, see Table 5-1).

The interviews took place in a private office and involved one interviewer. Interviewees were asked to recall and describe, in detail, the injury incident in which they were involved in previously, and the probes were used to interrogate the interviewee regarding the injury incident and the factors influencing their decision-making prior to the incident occurring. All interviews were recorded using a dictaphone and transcribed post interview using Microsoft Word.

5.4 ANALYSIS PROCEDURE AND RESOURCES INVESTED

For data analysis purposes, the CDM interview transcripts were transcribed and coded by one analyst. Coding of the transcripts involved identifying keywords from each interview question response, which in turn allowed key themes to be identified from the data. For example, for the response 'the packaging of the product and its signage were definitely the most influential' to the probe 'what was the most influential factor/ piece of information that influenced your decision-making at this point?', the keywords 'packaging', 'product' and 'signage' were extracted. This process of identifying keywords was iterative and involved reviewing the interview transcripts on multiple occasions. The key themes identified were then analysed using frequency counts.

The overall process was time consuming, with each CDM interview taking around 40 minutes to complete and two hours to transcribe. The interview coding procedure then took around five days to complete.

Table 5-1   CDM probes (adapted from O'Hare et al. 2000 and Crandall et al. 2000)

Goal specification	What were you aiming to accomplish through this activity?
Assessment	Suppose you were to describe the situation at this point to someone else. How would you summarise the situation?
Cue Identification	What features were you looking for when you formulated your decision? How did you know that you needed to make the decision? How did you know when to make the decision?
Expectancy	Were you expecting to make this sort of decision during the course of the event? Describe how this affected your decision-making process.
Options	What courses of action were available to you? Were there any other alternatives available to you other than the decision you made? How/why was the chosen option selected? Why were the other options rejected? Was there a rule that you were following at this point?
Influencing factors	What factors influenced your decision-making at this point? What was the most influential factor that influenced your decision-making at this point?
Situation Awareness	What information did you have available to you at the time of the decision?
Situation Assessment	Did you use all of the information available to you when formulating the decision? Was there any additional information that you might have used to assist in the formulation of the decision?
Information integration	What was the most important piece of information that you used to formulate the decision?
Experience	What specific training or experience was necessary or helpful in making this decision? Do you think further training is required to support decision-making for this task?
Mental models	Did you imagine the possible consequences of this action? Did you create some sort of picture in your head? Did you imagine the events and how they would unfold?
Decision-making	How much time pressure was involved in making the decision? How long did it actually take to make this decision?
Conceptual	Are there any situations in which your decision would have turned out differently?
Guidance	Did you seek any guidance at this point in the task/incident? Was guidance available?
Basis of choice	Do you think that you could develop a rule, based on your experience, which could assist another person to make the same decision successfully?
Analogy/ generalisation	Were you at any time, reminded of previous experiences in which a similar/different decision was made?
Interventions	What interventions do you think would prevent inappropriate decisions being made during similar incidents in the future?

5.5 OUTPUTS

An example CDM interview transcript is presented in Table 5-2, following which a summary of the CDM interview findings is presented in Table 5-3.

Table 5-2   Example CDM interview transcript

STORE	XXXXXXXXXXXXXX
ACTIVITY	Lifting large product (flatpack) onto trolley with customer
INCIDENT	*Basically a heavy box was involved, we're talking a flat pack, it was a large product which we no longer stock but at the time they were around the 50–60 kilo mark. There were two customers and myself, one of the customers had a trolley, the other customer was helping me lift it onto the trolley and his mate with the trolley was being a bit of an idiot and every time we'd go to put it down he'd pull away the trolley as a bit of a joke. What actually happened is the other guy who was actually physically helping me lift this thing up, he thinks that I'm closer to the trolley and he just lets go. Basically it's nowhere near the trolley and it's landed on top of me and I've gone to get up with this heavy implement on top of me and as I've come up I've twisted a little bit and its twinged my back, so yeah, it's a case of customers being stupid and not quite doing the right thing … erm … it was a busy day on a Saturday … erm … in hindsight, yes, it would have been a good idea to have another staff member to help me put it into the truck or whatever but you know, if you've got someone who is bigger and stronger than you offering to give you a hand, that was the thing at the time. I fell to the ground and this thing landed on top of me …*
Goal specification	**What were your specific goals at this point in time? What were you aiming to accomplish through this activity?** *Help the customer get this product through the register and pay for it and get him on his way.*
Cue identification	**What features were you looking for when you formulated your decision?** *On the day in particular we were very short staffed, extremely short staffed, and I made a wrong decision at the time, I suppose, in hindsight. I basically just, you know, the customer offered to help lift it up onto the trolley. Everything was fine when he was holding onto it and it only went pear shaped when he decided to drop it and couldn't be blowed hanging onto it any longer. The customer was another six inches taller than me and muscles on muscles, he probably looked like he could have lifted the thing by himself.*
Expectancy	**Were you expecting to make this sort of decision during the course of the event? Describe how this affected your decision-making process.** *– Well look, a lot of training has gone on since this has happened. The accident happened over two years ago … erm … basically these days I don't rely on customers at all, because they've proven to be a backwards step and now I don't rely on customers these days.*
Options	**What courses of action were available to you? Were there any other alternatives available to you other than the decision you made?** *1. Get a staff member, have a hostile customer who was in a hurry to get going with about another six or seven people waiting around for my services to help them where they were, so yeah, the pressure was on that day. Trying to get another staff member at the time because we were short staffed, we had a couple of sick leaves off … erm … I would have been waiting for quite a substantial time which would have been probably beyond the realms of how long the customer was prepared to wait.* **How/why was the chosen option selected?** *– Short staffed; customer pressure.*
Influencing factors	**What factors influenced your decision-making at this point?** *– Short staffed, I knew how busy it was, and customers were waiting.* **What was the most influential factor that influenced your decision-making at this point?** *– Customer was a little bit hostile, almost stand over tactics, 'Come on, mate, I haven't time to stuff around', effectively was his words he used, cos I did offer originally to get someone to help me lift it into the trolley and lift it onto their truck for them but he said, Come on, I can help you lift it', so …*

Table 5-3 Summary of CDM interview findings

CDM Probe	Summary of responses
What features were you looking for when you formulated your decision?	**55%** Product or elements associated with product, such as weight, packaging, size, shape **22%** Shelving, **22%** Task requirements, **22%** Desired placement of stock **18%** Experience of similar task, **18%** Availability of other workers for assistance **12%** Customer **8%** Availability of equipment **6%** Procedures **4%** Safety, **4%** Time pressure, **4%** Orders, **4%** Convenience
Were you expecting to make this decision during your shift?	**65%** Yes, **14%** No, **21%** No response or not sure
How did you know that you needed to make this decision?	**30%** Empty shelves **22%** Customer request **6%** Order from supervisor **2%** Request of fellow worker, **2%** Impending store visit from higher managerial staff
What courses of action were available to you? Were there any other alternatives available to you other than the decision you made?	**55%** Chosen course of action plus one other alternative course of action **18%** Chosen course of action plus two other alternative courses of action **18%** Chosen course of action was the only one available under the circumstances **8%** Chosen course of action plus three other alternative courses of action
What were the other courses of action available?	**37%** Use of appropriate equipment **20%** Get assistance of fellow worker, **20%** Use appropriate procedure **12%** Use appropriate lifting technique **12%** Other
How/why was the chosen option selected? Why were the other options rejected?	**22%** Limited availability of equipment **18%** Limited availability of fellow workers to assist, **18%** Customer pressure, **18%** It was the only course of action available **12%** Previous experience of performing the task in this manner **10%** Easiest way to perform the task/convenience **4%** Product in question, **4%** Time pressure
What factors influenced your decision-making at this point?	**44%** Product **20%** Customer **16%** Availability of fellow workers for assistance, **16%** Experience of similar task, **16%** Shelving **10%** Availability of equipment **6%** Convenience, **6%** Time pressure, **6%** Rules
What was the most influential factor that influenced your decision-making at this point?	**33%** Product **16%** Customer **12%** Availability of equipment **8%** Experience **6%** Shelving **4%** Availability of fellow workers for assistance, **4%** Time pressure, **4%** Rules, **4%** Convenience

Table 5-3 Concluded

CDM Probe	Summary of responses
What information did you have available to you at the time of the decision?	51% Product 18% Experience of task 14% Customer 10% Availability of fellow workers for assistance 8% Shelving, 8% Signage 6% Equipment availability
Did you use all of the information available to you when formulating the decision?	47% Yes, 6% No
Was there any additional information that you might have used to assist in the formulation of the decision?	39% No 4% Availability/location of other workers, 4% Warning signage 2% Time before store visit, 2% Availability/location of equipment
What specific training or experience was necessary or helpful in making this decision?	32% Experience of similar task was helpful 30% Manual handling training was helpful but not specific enough for this task 20% No training/experience was helpful 18% Manual handling training
Do you think further training is required to support decision-making for this task?	30% Yes, 48% No, 2% Revision of current manual handling training
Did you imagine the possible consequences of this action? Did you imagine events and how they would unfold?	84% Yes, 8% No
Was time pressure involved in making the decision?	53% Yes, 42% No
How long did it actually take to make this decision?	73% Decision was immediate 4% 10 minutes plus 2% Up to 10 minutes
Are there any situations in which your decision would have turned out differently? If so, what are these situations?	61% Yes 26% If fellow workers were available to assist 18% No 8% If appropriate equipment was available 4% If product was packaged more appropriately, 4% Less pressure from customer, 4% More support from supervisor/manager 12% Other
Did you seek any guidance at this point in the task/incident?	2% Yes, 83% No
Was guidance available?	53% Yes, 26% No, 2% Don't know
What interventions do you think would prevent inappropriate decisions being made during similar incidents in the future?	24% More equipment, 22% More workers, 18% Shelf redesign, 14% Training 4% Store design, 4% Personal protective equipment, 4% New procedures 38% Other

5.6 DISCUSSION

In safety critical domains, violations have previously been identified as behaviours which increase the likelihood of accidents occurring (Alper and Karsh 2009). Analogous to human error, it is generally accepted that the causes of violational behaviour exist both within and outside of individual workers, and often reside within the wider organisational system; however, little is known on what these causes actually are within specific contexts. Acknowledging that the organisation in question has an exemplary safety record, the aim of the study described was to identify the different factors which influence warehouse superstore workers' propensity to engage in violational behaviours. Coding of the interview transcripts identified pertinent characteristics associated with store worker decision-making prior to involvement in injury-causing incidents.

Regarding the information used when making decisions, the product, task, shelving, past experience, end location of product, customer, availability of other store workers, equipment, procedures, safety, time, orders and convenience were all cited as sources of information used to aid the decision-making process. Of these factors, the product involved was the most commonly reported factor (55 per cent), and the task requirements, shelving, and placement or desired location of the product involved were reported by over 20 per cent of the store workers interviewed.

The information used and available when making decisions was found to be important. Almost half of store workers felt that they had used all of the information available, whereas only 6 per cent felt that they had not. Over half of the store workers interviewed felt that further information would have been useful, including information regarding the availability and location of other store workers, more explicit warning signage on products (i.e. actual weight rather than advisory weight range), the exact time of higher management store visits, and information regarding the availability and location of equipment.

When questioned regarding the *factors* influencing decision-making, the product involved in the task was the most commonly reported influencing factor, with almost half identifying it as a factor influencing their decision-making. Following this, other influencing factors commonly identified by store workers included customers, other worker availability, shelving, experience of the same task, and equipment availability. When questioned what the *most* influential factor was, 33 per cent identified the product in question as the most influential factor. This was followed by the customer, equipment availability, and experience of the same task.

One important finding was that, when faced with a decision regarding which course of action to take for a particular task, almost three quarters of the interviewees (73 per cent) felt that they had alternative, more appropriate, courses of action available to them than the one they eventually chose (55 per cent felt that there was one alternative and 18 per cent felt that there were two alternative courses of action). Alternative courses of action cited included using the appropriate equipment, performing the task with assistance from another store worker, using the appropriate procedure, and using the appropriate lifting technique. When questioned why alternative, more appropriate, courses of action were not pursued, equipment and store worker availability, customer pressure, experience of doing the task successfully in another manner, and convenience were cited as key factors.

Visualisation or simulation of events as they might occur is a key factor in efficient decision-making (Klein et al. 1986). When asked if they considered the potential consequences of their chosen course of action, 84 per cent of the store workers interviewed reported that they did not consider the consequences. Further, 80 per cent did not picture events and how they might unfold.

The role of training and experience in decision-making was also explored. In the context of manual handling incidents, a third of store workers reported that, although they had had training in general manual handling skills, the training provided was not specific to different products or departments and so was not helpful in most cases. A third of store workers felt that additional manual handling training, focusing on specific products and departments, would be helpful in future. Regarding guidance for the task at hand, over three quarters of the store workers interviewed reported that they did not seek guidance, with only one reporting that they did seek guidance. Over half suggested that, in the event that they did seek guidance, appropriate guidance would have been available, whereas a quarter felt that appropriate guidance would not have been available at the time due to factors such as lack of knowledge, experience and poor attitudes toward safety.

Just over half of the interviewees suggested that time pressure was a significant factor in their decision-making, and almost three quarters (73 per cent) reported that the decision was made immediately with very little time incurred. This is important as it indicates that few store workers engage in a risk assessment process prior to making decisions, which is further evidenced by the lack of consideration of consequences or of how events might unfold.

Regarding situations in which their decisions would have turned out differently, almost two thirds (61 per cent) of the interviewees felt that there were situations in which they would have made a different decision. These situations included instances where other store workers were available to help, instances where appropriate equipment was available, instances where elements of the product (e.g. packing, signage) were different, instances where customer behaviour was more appropriate, and instances where more support was given from the coordinators involved.

Finally, the store workers were questioned over interventions that could be made that would prevent inappropriate decisions being made in future. The most commonly suggested interventions were more equipment, more store workers on the shop floor, redesign of the workspace (e.g. shelving), and additional training. Other popular interventions suggested included store redesign, and additional personal protective equipment.

The CDM analysis reveals various characteristics associated with store workers' decision-making and behaviour that require further attention. First, the data obtained suggests that the product in question is a key factor in store workers' decision-making, with it being identified by the majority of interviewees as the most important piece of information used, and the most influential factor involved in the decision-making process. This suggests that the products stocked at the stores have a key role to play in the enhancement of safe decision-making, with aspects such as size, shape, weight, signage and packaging all representing avenues for supporting appropriate decision-making and limiting risk. For example, when asked what further information would support more appropriate decision-making and behaviour, the interviewees commonly identified more explicit warning signage on the products, such as explicit actual weight warnings rather than two-person lift signage. Second, further information that could support the decision-making process but is not often available was identified; this included information on the availability of other store workers to help, improved warning signage on products, and information on the availability of equipment. Third, problems associated with the process of course of action selection by store workers were identified. It appears that, in the majority of cases, at least one other more appropriate, safer course of action is available to store workers, and that this is not pursued due to issues with equipment and other worker availability, customer pressure, experience of performing the task in a different manner and convenience. Store workers also do not appear to be considering the potential consequences associated with their chosen course of action, or picturing how events might potentially unfold. This is further evidenced by the fact that the majority of the decisions discussed were made 'immediately' with little or no time spent. Simulation of how courses of action might unfold has previously been identified as a key element in the decision-making process (e.g. Klein et al. 1986), thus these findings are especially problematic. The CDM responses with regard to course of action selection, consequence consideration and time taken to make decisions all point to a lack of engagement in any form of risk assessment process prior to making decisions regarding how to undertake a particular task. Fourth, the CDM findings suggest that the current manual handling training programme may not be sufficiently specific to the products and tasks involved in store workers' day-to-day activities, with many of the incidents involving manual handling tasks and products not covered in the current manual handling training programme. This suggests that more departmental and product-specific manual handling training programmes may be worthwhile, rather than merely general manual handling training. Fifth and finally, the CDM analysis indicates that in many cases there are situations in which store workers' decisions to engage in risky behaviour could have turned out differently. These include situations where other store workers were available to help, if appropriate equipment was available, if aspects of the product involved were more appropriately designed (e.g. packaging, signage), if customer behaviour was more appropriate, and if more support was given by the supervisors involved.

At a non-domain specific level, the findings of the present study provide further evidence for the notion that violations are indicative of system-wide problems rather than being exclusively actions made by 'bad' workers (e.g. Alper and Karsh 2009). Alper and Karsh (2009), e.g. conducted a review of 13 articles focusing on the causes of violations in healthcare, aviation, mining, rail, commercial truck driving, and construction. In conclusion, they identified 57 causes or predictors of violations within the following categories: individual characteristics, information, education and training, design to support worker needs, safety climate, competing goals, and problems with rules. The findings from the present study corroborate the conclusions of Alper and Karsh (2009), who concluded that the causes of violations are present across organisations, and also that they are likely to interact to create situations in which individuals are more likely to violate procedures. Of the 57 predictors identified by Alper and Karsh (2009), 29 were identified in the present study, with the majority of those not identified in the present study being domain specific (e.g. medical class, hours driven per week, expectation by doctor, airline type).

Based on a synthesis of the literature focusing on worker engagement in risky behaviours from similar domains (e.g. Choudry and Fang 2008; Lombardi et al. 2009), personal, product, management, equipment, environmental, safety culture-related, time pressure and training factors have all been found previously to have an influence on worker engagement in unsafe or risky behaviours (e.g. Denis et al. 2006; Choudry and Fang 2008; Lombardi et al. 2009; St-Vincent et al. 2005). Factors from each of these different categories were found to influence store worker decision-making and behaviour in the present study. Focusing on research undertaken previously with a specific focus on the warehouse superstore domain, St-Vincent et al. (2005) classified the factors influencing manual handling activities in warehouse superstores into the following categories: workplace layout, products, equipment, and stock management. The findings from the present study confirm this, with similar product and equipment categories of influencing factors identified, and also workplace layout and stock management factors being covered in the products and environmental factor categories. Specifically, within the workplace layout category, St-Vincent et al. reported that marketing strategies, display characteristics and space limitations affected handling activities. Similarly, product placement, display requirements, and space limitations were found to be key influencing factors in the present study. Within the products category St-Vincent et al. (2005) identified physical and packaging characteristics, and visual presentation requirements as factors influencing risk. These three factors were also found to be prominent influencing factors in the present study, particularly the physical and packaging characteristics of the products sold. The equipment factors identified in the present study were also similar to those found by St-Vincent and colleagues, with pallet jack maintenance and inappropriately designed equipment being identified in both studies. Finally, stock management issues similar across both studies included poor communication and planning and inadequate information being given to stockers.

5.7 SUMMARY OF KEY FINDINGS

It is concluded from CDM interviews that, in this domain at least, the factors influencing store worker decisions to engage in violational behaviours are multi-dimensional, and exist across all levels of the organization. They range from factors related to the individuals involved and the equipment that they use, to supervisory/management and wider organisational factors at the higher company management levels. In the stores analysed, store worker decision-making is currently influenced by a range of factors present at all levels of the organisation. These include task, personal, customer, product, equipment, environmental and organisational factors. This is in line with the prominent systems-based models of error and accident causation (e.g. Rasmussen 1997; Reason 1990), since influencing factors were identified across all levels of the organisation, including the higher managerial levels.

5.8 ACKNOWLEDGEMENTS

The authors would like to acknowledge the retail organisation involved for funding this research and all of the personnel who made this research possible through participation in the data collection activities undertaken. We would also like to acknowledge Margaret Trotter and Elizabeth Varvaris from the Human Factors Group at the Monash University Accident Research Centre for all of their efforts in reviewing the literature and data transcription and analysis.

6
Propositional Networks: Challenger II Tank Friendly Fire Case Study

6.1 INTRODUCTION

Fratricide is defined by the US Army as: 'the employment of friendly weapons and munitions with the intent to kill the enemy or destroy his equipment or facilities, which results in unforeseen and unintentional death or injury to friendly personnel' (US Army, cited in Wilson et al. 2007). Commonly referred to as 'friendly fire', the problem currently represents a major issue within military systems worldwide (Rafferty et al. 2010; Wilson et al. 2007). For example, since 2001, at the time of writing this book, up to eight British Armed Forces personnel had been killed in confirmed and suspected friendly fire incidents in Afghanistan and Iraq (BBC News 2011). Further, during the first Gulf War conflict, 35 (24 per cent) of the 146 casualties suffered by US Forces were caused by friendly fire incidents (Ripley 2003) and nine (38 per cent) of the 24 casualties suffered by British Forces were attributed to friendly fire (Cooper 2003).

One factor commonly cited as contributing to friendly fire incidents is poor situation awareness on behalf of those involved (e.g. Rafferty et al. 2010; Gorman et al. 2006). Although the involvement of poor situation awareness in friendly fire incidents is somewhat obvious, since friendly force personnel are often wrongly assumed to be enemy force personnel, a precise description of exactly how situation awareness fails during these incidents is yet to be offered. The case study presented in this chapter involved the use of propositional networks (Salmon et al. 2009; Stanton et al. 2006) to describe and analyse situation awareness during a recent high-profile friendly fire incident. The exploratory analysis was undertaken on the premise that problems leading to friendly fire incidents are directly linked to the concept of SA. As such, the aim of the analysis was to identify the different types of SA-related failures involved in the friendly fire incident with a view to informing future prevention efforts.

6.2 INCIDENT DESCRIPTION

The Challenger II tank friendly fire incident, in which a UK Armed Forces Challenger II tank opened fire on two fellow UK Armed Forces Challenger II tanks, killing two occupants, was the focus of this case study. The following description of the incident was adapted from that presented in the UK Ministry of Defence (MoD) Army board of inquiry report (MoD 2004).

As part of Operation TELIC the 1st battalion the Royal Regiment of Fusiliers (1 RRF) and the 1st battalion the Black Watch (1 BW) formed two of the Battle Groups (BGs) of the 7th Armoured Brigade. Attached to the 1 RRF were C Squadron, Queens Royal Lancers (C Sqn QRL) and attached to the 1 BW were Egypt Squadron, 2nd Royal Tank Regiment (2 RTR). The organisational structure of the forces involved is represented in Figure 6-1.

In March 2003 the 7th Armoured Brigade were engaged in operations concerning key bridges over the Shatt al Basra canal in the western outskirts of Basra (MoD 2004). Upon seizure of two bridges (hereafter referred to as bridges A and B), the BGs placed themselves on the bank of the canal in order to deny enemy approaches. On 24 March 2003, 1 RRF handed over bridge B to 1 BW, who subsequently placed themselves in a compound on the western side of the canal in co-location with Egypt Sqn 2 RTR. 1 BW deployed a force on the Basra side of the canal consisting of a warrior (WR) platoon and a tank troop of Egypt Sqn 2 RTR. Also on 24 March 2003 the boundaries of C Sqn QRL were expanded to include responsibility for a dam located approximately 1,400 metres to the north of bridge B.

In the early hours of 25 March 2003 two Challenger II tanks from the C Sqn QRL were located in a support position adjacent to the dam. At the same time, 1BW BG were defending a bridge over the canal approximately 1,500 metres to the south-east of the Challenger II tanks position. Unaware of the friendly forces located in the vicinity of the dam, the commander of a 2 RTR Challenger II tank, upon sighting hot spots through his thermal imaging equipment, wrongly classified them as enemy personnel entering and vacating a bunker (the hotspots were actually the C Sqn QRL Challenger tanks located adjacent to the dam). Upon requesting, and being granted, permission to engage the hotspots the 2 RTR tank fired two High Explosive Squash Head (HESH) rounds at one of the C Sqn QRL Challenger II tanks. The first round landed short but the effects of the blast threw the crewmen from the tank turrets. Six minutes after the first round was fired, the 2 RTR observed and classified a moving object as an enemy armoured vehicle and fired the second HESH round which was a direct hit, detonating in the commander's hatch killing its two occupants. Two other crew members received serious burns and other injuries.

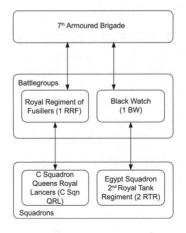

Figure 6-1   Hierarchical organisation of the system

The official UK Army inquiry into the incident identified various causal factors, including planning, oversights, communications failures, failures in command and control, lack of SA, incorrect target identification, inadequate tactics, techniques and procedures, fog of war, and fatigue. Specifically, the inquiry found shortcomings in the following areas:

1. The passage of information regarding inter-BG boundaries, locations of friendly forces, and the existence of a key tactical feature;
2. Command and control issues including co-ordination of the battlespace, co-operation and unity of effort; and
3. Combat identification, including the issues of SA, target identification and procedures for the handover of a tactical position, including briefings, allocation of arcs of fire and designated reference points.

The shortcomings identified contributed to misunderstandings about the correct arcs and enemy threat which led to an erroneous appreciation of the situation. This led to 2RTR engaging the targets observed. The board concluded that the incident could have been prevented if the following had occurred (adapted from board of inquiry report):

1. Complete and accurate briefings had been given regarding the boundaries in place and also more timely dissemination of the Bde trace (dated March 2003);
2. OC B Coy's orders had included the assumption made about the Company boundaries as well as the existence and tactical significance of the dam to operations on bridge B;
3. All troops had received a formal tactical brief for their task in the compound prior to deployment over the bridge following OC B Coy's orders. This brief should have included the existence of the QRL tanks at the dam when it became known;
4. Challenger tank Commander (2 RTR) had been more inquisitive regarding the details of the task over the bridge and had not accepted the lack of critical information;
5. The troop and platoon had acted in a co-ordinated and unified manner regarding arcs of fire, target reference points and enemy threat within a properly structured brief for handover of a tactical position;
6. Challenger tank Commander (2 RTR) had not been disorientated and placed the potential target on the wrong side of the canal;
7. There had been positive identification of the hotspots as enemy.

The various failures that contributed to this incident can all, in some way, be related to the level of situation awareness held by the overall military system throughout the scenario. Accordingly, the aim of the analysis presented was to identify the specific SA-related failures involved in the incident.

6.3 DATA SOURCES AND DATA COLLECTION

The primary data source was the official report derived from the UK MoD's Army board of inquiry report into the incident (MoD 2004). Other data, including news reports and articles (e.g. BBC website news articles) were also reviewed for contextual information.

6.4 ANALYSIS PROCEDURE AND RESOURCES INVESTED

The propositional network methodology was used to model situation awareness throughout the scenario, with a view to identifying the situation awareness failures that contributed to the incident. One analyst with significant experience in the propositional network method constructed the network based on content analysis of the incident description presented within UK MoD's Army board of inquiry report (MoD 2004).

6.5 OUTPUTS

An overall propositional network representing situation awareness throughout the scenario is presented in Figure 6-2. The situation awareness failures involved in the incident are then represented in Figure 6-3, where the circled information elements represent information elements that were involved in an situation awareness failure of some sort.

From the propositional network description it is possible to identify the following failures related to the information elements underpinning the system's situation awareness that contributed to the incident:

1. False information elements. These represent information elements that formed part of the system's situation awareness that were in fact false, i.e. were not present in the actual situation. False information elements in this case included the enemy, the enemy ammunitions dump and the enemy bunker.
2. Inaccurate/misunderstood information elements. Include information elements that formed part of the system's situation awareness but that were inaccurate or misunderstood, including the enemy's location, the boundaries, the locations of friendly and enemy forces, the Challenger II tank commander's disorientation, the erroneous target classification, and the confusion over the arcs of fire in force.
3. Missing information elements. This class of information elements refers to information that should have formed part of the system's situation awareness but for various reasons did not. Key missing information elements in this case included the dam, which the Bde staff were not aware of and the presence of other friendly forces in the area.
4. Non-communicated information elements. A number of different information elements required were known by parts of the system but were not communicated to other parts. These include the dam and the presence of the friendly Challenger II tanks at the dam.

The four classes of information element failures identified all played a part in the incident. It is also possible from the content analysis to identify the causes of the four types of information element failures. For example, processes that were inadequately performed were found to be particularly problematic. Examples include the planning oversight which led to a lack of awareness of the dam. Further, inadequate tools were also found to be problematic. These included tools that were used by those involved that performed inadequately, such as the maps and aerial photography used during the planning stages.

CHAPTER 6 • PROPOSITIONAL NETWORKS: CHALLENGER II TANK FRIENDLY FIRE CASE STUDY

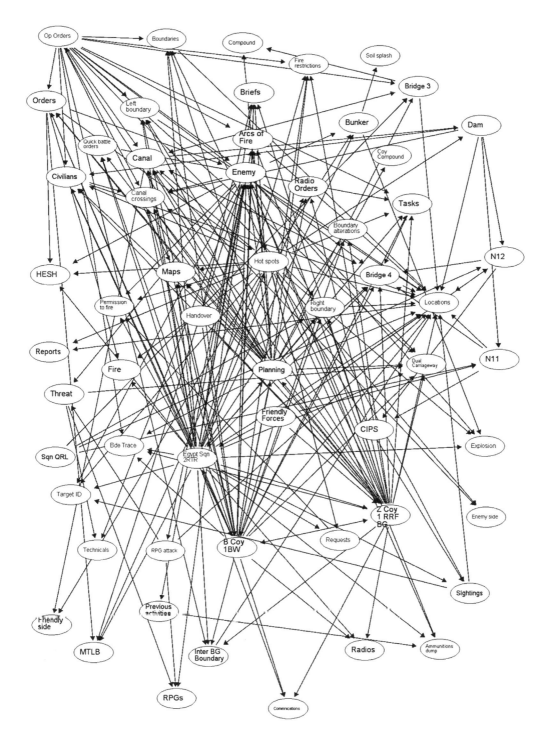

Figure 6-2   Overall incident propositional network

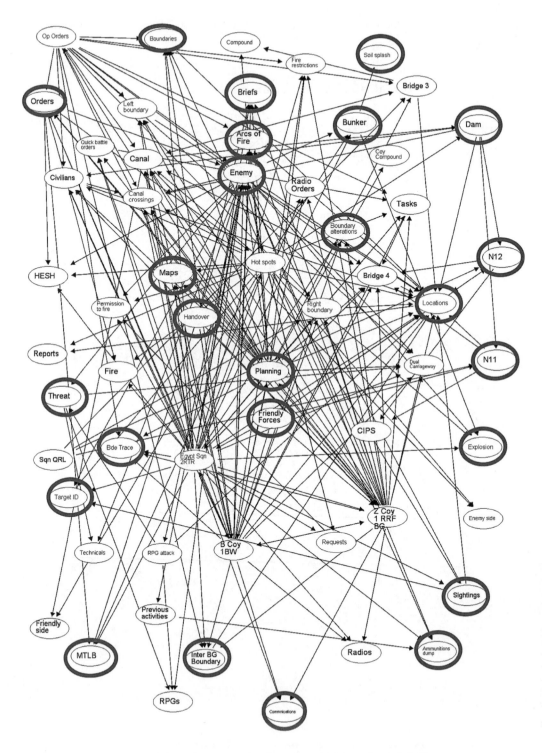

Figure 6-3   Propositional network showing DSA-related failures

The information element failures along with the causal elements are presented in Table 6-1.

Table 6-1   Information element failure types and causal factors

False information Elements	Incorrect Information Elements	Misunderstood Information Elements	Missing Information Elements	Uncommunicated Information Elements	Inadequate Process	Inadequate Tools
Enemy	Location — Enemy	Enemy — Location	Dam	Dam	Planning	Maps
Bunker	Enemy	Friendly C2s — Threat	Friendly C2s	Friendly C2s	Comms	Arial photography
Ammo dump	Orientation	Boundaries	Boundary alterations	Boundary alterations	Orders	Range card
	Target ID	Arcs of fire — Fire Control	Hotspot grid reference		Handover	
	Threat assessment	Target ID				

From a situation awareness perspective then, the following failures were identified:

1. Lack of awareness of the dam. During the planning process, the Bde staff were not made aware of the dam and the resources used during the planning process did not contain information regarding it. For example, the dam was not present on the aerial photography of the area and it also did not appear on the various scale maps that were used (MoD 2004). Further, although the dam could be seen from the bridge, most of the troops passing over it had only limited observation. The OC B Coy was aware of the dam but considered it to be out of his boundaries and so did not impact his operations on Bridge 4. For this reason none of the OC B Coy's plans took the dam into account (MoD 2004).
2. Lack of awareness of the boundaries in force. The change of boundaries between 1 RRF and OC B Coy 1 BW was not communicated effectively to B Coy 1 BW (MoD 2004). As a result of this both OC Z Coy and OC B Coy 1 BW were not aware of the details of the boundary changes. Also, the OC B Coy's orders (given to his crew commanders) did not explain his assumptions over the left boundary and the reason for imposing the western edge of the compound as his left of arc (MoD 2004).
3. Lack of awareness of other friendly forces in the area. The troops located in the compound were not aware of the QRL tanks located at the dam (MoD 2004). The inquiry concluded that the locations of the QRL tanks had not been marked on the central company operations map, recorded in the radio log or disseminated with the rigour that was required to ensure the necessary acknowledgement. Prior to the incident Challenger tank Commander (2RTR) had not been able to establish communications with either OC 5 PI or the Coy HQ but had good communications with Egypt Sqn HQ. Challenger tank Commander 2 RTR had spoken to Coy HQ

requesting information about locations of friendly forces and to clarify the inter BG boundary on the left (MoD 2004). This request, although passed up to BG HQ was not passed to B Coy HQ in the compound which was co-ordinating the battle space and in command of the armoured troop deployed over the canal. (The B Coy HQ therefore should have been the HQ to co-ordinate a response and assist the Challenger tank Commander in his decision-making process). However, the Coy HQ could not be contacted at the time at which time the Lieutenant in question had given no clear indication in his radio message that he was intending to engage a target (MoD 2004).

4. Challenger tank Commander 2 RTR's lack of spatial (location) awareness. There was an inaccurate orientation of the map in relation to the Gun Position Indicator (GPI) and the range return from the Laser Range Finder (MoD 2004). The lieutenant never considered that he was firing back across the canal, despite there being some easily recognisable features to aid accurate orientation (MoD 2004). There was a distance of 300 metres between the supposed enemy location and the actual location of the two QRL tanks. According to the inquiry report, this disorientation should have been remedied within the 30 minutes that elapsed between initial observation of the target and when the tank opened fire (MoD 2004).

5. Erroneous assessment of enemy threat. Challenger tank Commander 2RTR states that he was briefed that the enemy threat was from the left flank (MoD 2004). He was expecting to see dismounted RPG (Rocket Propelled Grenade Launcher) teams in this area and when the hotspots were observed they conformed to what he asserts he was briefed on but outside a range that posed an immediate threat (MoD 2004). According to the inquiry report, the identification of enemy far over to the left flank should have warranted more rigorous interrogation and corroboration (MoD 2004).

6. Erroneous Target Identification. Initially Challenger tank Commander RTR observed hotspots which he considered were people moving about on top of a bunker, but these were discounted as being civilians (MoD 2004). According to the inquiry report, 'the crew did not complete the target identification process on positively identifying the hotspots as enemy' (MoD 2004). Also, following the first round of firing an armoured vehicle was incorrectly identified (MoD 2004).

7. Confusion of fire control and discipline, arcs of fire and target ID. There was a failure to undertake a detailed handing over of the tactical position which led to confusion over issues such as fire control and discipline, arcs of fire and target ID (MoD 2004). A range card was neither handed to, nor used by, Challenger tank Commander 2RTR. The inquiry report stated that, 'it is the opinion of the board that the lack of situation awareness of Challenger tank Commander 2 RTR could have been ameliorated by the handing over of a comprehensive range card' (MoD 2004). There was a misunderstanding by the crew of Challenger tank Commander 2 RTR as to the arcs of fire and the enemy situation and there was also confusion over the exact target ID that was given (MoD 2004). Finally, the reports of the hotspots from Challenger tank Commander 2 RTR sent over the Company radio net did not include a grid reference. The board of inquiry concluded that the lack of a grid reference led to a misunderstanding regarding where the alleged enemy was actually located, which made accurate target identification impossible (MoD 2004).

6.6 DISCUSSION

The purpose of this case study was to use the propositional network method to identify and represent the situation awareness related failures involved in the Challenger II tank military friendly fire incident. The analysis indicates that the propositional network approach lends itself to the analysis of accidents occurring within complex collaborative systems, particularly when situation awareness failures are involved. In this case, there were various failures in the military system's situation awareness development and maintenance that contributed to the tragic incident. These failures were categorised into the following types: instances where false (erroneous) information underpinned the system's SA; instances where information elements were misunderstood by different components of the system; instances where critical information was missing from the system's SA; instances where information critical for situation awareness was not communicated to the appropriate component of the system; and instances where inadequate processes and tools impacted the situation awareness achieved by the system.

The misidentification of the enemy target and the resultant engagement, although in the immediate term was actually caused by the Challenger II tank commander's erroneous classification of the friendly tank as enemy personnel, was caused by various failures, all of which can be linked to the situation awareness held by the military system as a whole. For example, the lack of awareness of the dam was caused by a planning oversight and the use of inadequate mapping and aerial photography. Various other key pieces of information critical for accurate situation awareness were not adequately communicated around the system, including the boundaries and their subsequent alterations and the location of the QRL Challenger II tanks at the dam. Various flaws were also identified in the command and control process adopted, including the lack of unity of effort on the bridge and the lack of co-ordination between Coy HQs. Finally, a lack of spatial awareness on the part of the 2 RTR Challenger II tank commander was also involved.

6.7 SUMMARY OF KEY FINDINGS

The propositional network analysis revealed a number of SA-related failures that contributed to the friendly fire incident. These failures were categorised into the following groups:

1. False information elements. A number of the information elements underpinning the system's situation awareness were in fact false, i.e. not present in the actual situation. For example, there was no enemy present, neither was there an enemy ammunitions dump or bunker.
2. Inaccurate/misunderstood information elements. The system's understanding of various information elements critical to situation awareness was found to be inaccurate, including the enemy's location, the boundaries, the locations of friendly and enemy forces, the target classification, and the confusion over the arcs of fire in force.
3. Missing information elements. Various information elements that were required for accurate situation awareness during the scenario were found to be absent from different components of the systems situation awareness. For example, the Bde staff were not aware of the dam or of the presence of other friendly forces in the area.

4. Non-communicated information elements. A number of the different information elements required were known by parts of the system but were not communicated to other parts. These include the dam and the presence of the friendly Challenger II tanks at the dam.

It was concluded from the analysis that the four classes of SA-related failures all played a part in the incident.

6.8 ACKNOWLEDGEMENT

This work from the Human Factors Integration Defence Technology Centre was part-funded by the Human Sciences Domain of the UK Ministry of Defence Scientific Research Programme.

7
Critical Path Analysis: Ladbroke Grove Case Study

7.1 INTRODUCTION

On 5 October 1999 at 8.06 am, a light turbo commuter train travelling from London to Bedwyn (Wiltshire in the UK) collided with a high-speed train travelling in the opposite direction just three minutes after departing Paddington station. The collision occurred at a combined speed of 130 mph, killing both train drivers and 29 passengers and injuring over 400 passengers. Although there were many failures that contributed to the incident (Lawton and Ward 2005), one of the key issues explored in the subsequent inquiry (Cullen 2000) was why the signaller involved took 18 seconds to respond to the alarm advising him of an unauthorised track occupation. In his inquiry report, Lord Cullen criticised the time taken by the signaller to respond, whilst conceding that an earlier response would not have prevented the crash from occurring (Cullen 2000).

Known colloquially as the Ladbroke Grove incident, the collision has previously been analysed using Reason's Swiss cheese model of accident causation (Lawton and Ward 2005). Lawton and Ward's analysis identified a number of failures, including train driver (e.g. passing red signal) and signaller failures (e.g. failure to respond appropriately) and failures across the wider organisational system, including system design failures (e.g. track complexity), inadequate defences (e.g. lack of safety devices, poor signalling, and organisational failures (e.g. poor safety management, inadequate training). The purpose of the analysis presented in this case study, however, is to determine whether the signaller's response time, criticised in the inquiry report and also highlighted in Lawton and Ward's analysis, was indeed inappropriate. For this purpose, the Critical Path Analysis (CPA) approach was used to model the response times of the signaller. The purpose of the analysis was to offer an independent judgment regarding how long it could reasonably be expected for the signaller to respond in this context.

7.2 INCIDENT DESCRIPTION

A schematic representation of the Ladbroke Grove incident is presented in Figure 7-1. A light commuter train (represented by the headcode IK20 on the signaller's screen, hereafter referred to as train one) had erroneously passed a red signal labelled SN109 on the track out of Paddington station as if signal SN109 was set to green, indicating that the train may proceed safely. A high speed train (represented by the headcode 1A06 on the signaller's screen, hereafter referred to as train two) was approaching Paddington from the opposite direction (see arrows depicting both trains' direction of travel in Figure 7-1). The Ladbroke Grove rail junction has six tracks, three of which are illustrated in Figure 7-1. The tracks are divided into segments, called track circuits, which indicate the position of the train. Train one travelled through track circuits GD, GE (passing signal 109 – the red signal), GF and GG as indicated by the arrow going left in Figure 7-1, whilst train two travelled through track circuits MX, MY and MZ as indicated by the arrow going right in Figure 7-1.

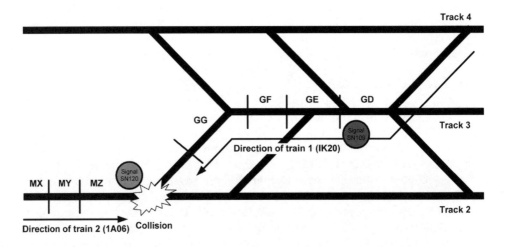

Figure 7-1   Ladbroke Grove incident track layout schematic

A timeline of the events involved in the incident, describing the information that would have been available to the signaller, is presented below (adapted from Lord Cullen's inquiry report). The timeline picks up the journey from the moment that train passed signal SN109, which was set to red. The train driver should have waited at this signal until it turned green.

08:08:29 – Track circuit GE occupied IK20 warning message is presented on the alarm screen in the signalman's work station and auditory 'tweet' sounds (track occupation alarm referring to track circuit GE by train one) – at the same time a red line appears on the track layout on the track display and the train headcode of IK20 stays at signal 109, the red signal.

08:08:32 – The oncoming train two occupies track circuit MZ and a red line appears on the appropriate track display with the headcode 1A06 (the number associated with train two).

08:08:34 – Auditory alarm 'tweet' sounds as rear of train one clears track circuit GD (i.e. the track circuit before GE) and the track circuit is shown as cleared on track display.

08:08:36 – Track circuit GF occupied message displayed and auditory 'tweet' sounds (track occupation alarm referring to the occupation of track circuit GF by train one – at the same time a red line appears on the track layout on the track display).

08:08:41 – Rear of train one clears track circuit GE (i.e. the track circuit before GF) and track circuit shows as cleared on track display.

CHAPTER 7 • CRITICAL PATH ANALYSIS: LADBROKE GROVE CASE STUDY   145

08:08:42 – The rear of the oncoming train two clears track circuit MY and track circuit MY shows as cleared on track display.

08:08:49 – Track circuit GG occupied by IK20 alarm message displayed and auditory 'tweet' sounds referring to track circuit GG by train one – at the same time a red line appears on the track layout on the track display.

08:08:50 – Train one and train two collide.

The signaller's workstation is presented in Figure 7-2. The workstation comprises six screens, a trackerball and buttons, a keyboard, and four telephones. The track displays can be seen on the four screens from far right. These are similar to the track schematic shown in Figure 7-1, only with the complexity to reflect six tracks and the interconnections.

Unless the signaller is looking directly at the appropriate point on the alarm screen, the first notification of a new alarm is an auditory warning. There are four categories of alarm which are colour coded yellow, blue, green and red, presented in that order from top to bottom of the screen. Only those alarms that apply are displayed. All four categories have the same auditory 'tweet'. The track occupation alarms are colour coded in red and appear at the bottom of the screen. When the signaller had read the track occupation alarm, he would be aware that a train had overshot the point at which it was supposed to stop.

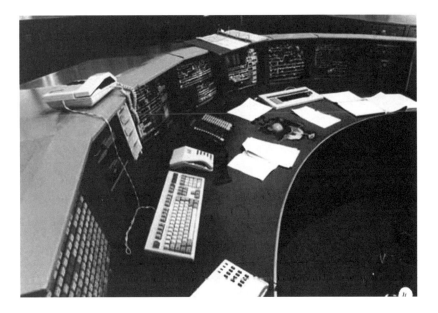

Figure 7-2   Signaller's work station

There are various reasons why a train might overshoot its stopping point. For example, the track occupation alarm can sound when trains are shunting (e.g. the manoeuvring of carriages and train cabs along the track and into sidings). This represents a false alarm, and the signaller would generally be expecting the alarm and thus will be able to acknowledge and ignore it. There are also a few instances in which the track occupation alarm is activated due to slight incompatibilities in the track circuit design and the direction of train running. This means that occasional false alarms are displayed and the signaller will be able to acknowledge the alarm and ignore it. Also, on some occasions, train drivers simply misjudge the stopping distance required. Most tracks have a safe overshoot area to cope with this; however it still represents a genuine alarm, and when it happens the driver involved normally calls the signaller once the train has come to a halt. For example, in the case being studied here, the signaller is likely to have had the expectation that the driver of train one would pull up within track segment GE in Figure 7-1. In fact, if the driver of train one had stopped at any point up to and including track segment GG, the accident would have been avoided. Given this background information, it is possible that the signaller (on noting an overshoot) could have sought confirmation of the event, i.e. waited for an additional alarm in order to confirm that this was a 'real' train runaway event.

On only a very small number of occasions will the track occupation alarm refer to a real train runaway in which a train has continued to run on a line for which it has not been cleared. This represents a real emergency; however, these events are rare and it is possible for a signaller to have never previously encountered a real overshoot event. The effects of the low Positive Predictive Value (PPV) of the alarm system have been shown to slow the human reaction times down considerably, particularly below a PPV of 0.25 (Getty et al. 1995). If this does happen, the signaller has to first find the track occupation on one of his four track displays and then decide what course of action to take. If the signaller decides that this is a real case of a runaway train, then either a stop message can be sent to the train or the points can be switched to send the train onto an unoccupied piece of track.

Both of these decisions, i.e. sending a stop message or switching the points, require the signaller to read the track ahead of the train in order to determine if there is an oncoming train and if either of the trains can be stopped in time. Sending a stop message would require the assessment of the stopping distance and the trajectory of the train(s). Moving the track points ahead of the train could divert the train onto track four instead of track two (see Figure 7-1) to prevent an impending collision, but the points can lock up when they detect a train is in the circuit in order to prevent an inadvertent point change. These complex decisions have to be made by the signaller who will be under considerable time-critical stress. The signallers' rule book does not give unequivocal guidance about which is the best strategy, it just requires the signaller to 'act immediately'. This begs the question to be addressed by the analysis presented in this chapter, could the signaller have responded quicker than 18 seconds?

7.3 DATA SOURCES AND DATA COLLECTION

A range of data sources were utilised for this analysis. For descriptive and contextual information regarding the incident, Lord Cullen's inquiry report was used. The CPA analysis process is supported by data on human operator task completion times. For this case study task timings were derived from the general Human-computer Interaction (HCI) literature. A summary of the task timings used are presented in Table 7-1.

CHAPTER 7 • CRITICAL PATH ANALYSIS: LADBROKE GROVE CASE STUDY

Table 7-1   Estimates of activity times from the literature on HCI

Activity	RT (ms)	Source
Read (alarm message, headcode, etc.)		
Read simple information	340	Baber and Mellor (2001)
Read short textual descriptions	1800	John and Newell (1990)
Recognise familiar words or objects	314–340	Olsen and Olsen (1990)
Hear (auditory warning)	300	Graham (1999)
Search (screen for alarm or train(s))		
Checking or monitoring or searching	2700	Baber and Mellor (2001)
Scanning, storing and retrieving	2300–4600	Olsen and Olsen (1990)
Primed search	1300–3600	estimated
Diagnosis or decision		
Mental preparation for response	1350	Card et al. (1983)
Choosing between alternative responses	1760	John and Newell (1990)
Simple problem solving	990	Olsen and Nielson (1988)
Speak (e.g.: 'We've got a SPAD')	100 per phoneme 1112	Hone and Baber (2001) Average time from speaking the phrase 10 times
Move hand to trackerball or keyboard	214–400 320	Card et al. (1983) Baber and Mellor (2001)
Move trackerball to target item Move cursor via trackerball 100mm	1500 1245	Olsen and Olsen (1990) Baber and Mellor (2001)
Press key (e.g. ACK or CANCEL key)	200 80–750 230	Baber and Mellor (2001) Card et al. (1983) Olsen and Olsen (1990)
Type headcode		
Average typist (40 wpm)	280	Card et al. (1983)
Typing random letters	500	Card et al. (1983)
Typing complex codes	750	Card et al. (1983)
Auditory processing (e.g. speech)	2300	Olsen and Olsen (1990)
Switch attention from one part of a visual display to another	320	Olsen and Olsen (1990)

7.4 ANALYSIS PROCEDURE AND RESOURCES INVESTED

The CPA modelling of the signaller's response times in the events from the presentation of the first SPAD alarm was undertaken based upon a method initially developed by Gray et al. (1993) and further refined by Baber and Mellor (2001). The analysis involved the following steps:

1. Analyse tasks to be modelled. The tasks need to be analysed in fine detail if they are to be modelled by multimodal CPA. Hierarchical tasks analysis can be used, but it needs to be conducted down to the level of individual task units. This fine grain level of analysis is essential if reasonable predictions of response times are to be made.

2. Allocate sub-tasks to input/processing/output modality. Each unit task then needs to be assigned to a modality. For the purposes of control room tasks, these modalities are as follows:

   a) Visual tasks: e.g. looking at the SCADA track displays, looking at alarm screen, and looking at written notes and procedures.
   b) Auditory tasks: e.g. listening for an auditory warning or listening to a verbal request.
   c) Central processing tasks: e.g. making decisions about whether or not to intervene and selecting intervention strategies.
   d) Manual tasks: e.g. typing codes on the keyboard, pressing button, and moving the cursor with the trackerball.
   e) Verbal tasks: e.g. talking on the phone, talking to another signalman in the control room.

3. Sequence the sub-tasks in a multimodal CPA diagram. The tasks are put into the order of occurrence, checking the logic for parallel and serial tasks. For serial tasks, the logical sequence is determined by the task analysis. For parallel tasks, the modality determines their placement in the representation. To make the CPA easier to view, tasks of the same modality are always positioned in the same row in the diagram, as shown in Figure 7-3.

4. Allocate timings to the sub-tasks: Timings for the tasks are derived from a number of sources. For the purposes of this exercise the timings used are based on the HCI literature, and are presented in Table 7-1. Standard times are taken to represent elemental human performance. These times are based on underlying aspects of human performance which is independent of context.

5. Determine the time to perform the whole task: The time that the task may be performed can be found by tracing through the CPA using the longest node-to-node values.

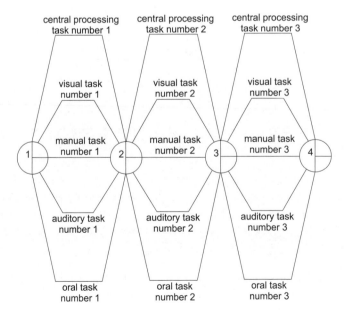

Figure 7-3   CPA representation

CHAPTER 7 • CRITICAL PATH ANALYSIS: LADBROKE GROVE CASE STUDY

7.5 OUTPUTS

The CPA analysis of the signalman's activities is presented in Figure 7-4. The timings are presented in milliseconds (ms).

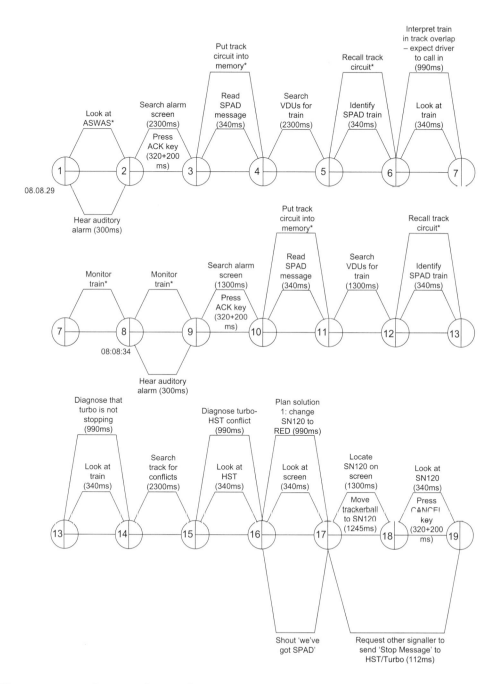

Figure 7-4 CPA analysis of signalman's activities post SPAD

7.6 DISCUSSION

Based upon the CPA, the predicted time to undertake the activities was 17.54s. This analysis support the evidence that the signalman took 18 seconds to change signal SN120 from green to red in response to the alarms associated with the runaway train during the Ladbroke Grove incident. The timings used for the analysis were based on published data of human response times in particular activities, with the exception of the vocalization of 'We've got a SPAD'. The low probability of the SPAD alarm relating to an actual SPAD (cf. Getty et al. 1995) might explain why the signalman's initial diagnosis assumed that the driver of the turbo train would come to a halt after passing signal SN109 and hence not consider addressing the points until later. All previous experiences of the track occupation alarm had shown this to be the case. It is likely, therefore, that the signalman revised this diagnosis on receipt of the second alarm.

In conclusion, the analysis presented therefore suggests that the signalman's response time of approximately 18 seconds was reasonable under the circumstances. This case study demonstrates how, during accident analysis efforts, time-based modelling can be used to evaluate the reaction time of human operators in emergency or accident scenarios.

7.7 SUMMARY OF FINDINGS

Based on the use of CPA to model the signaller's response time during the Ladbroke Grove incident, it is concluded that the previously criticised response time of 18 seconds was in fact reasonable under the circumstances.

7.8 ACKNOWLEDGEMENT

The authors would like to acknowledge Professor Chris Baber for his contribution to the following journal article on which this chapter is based:

Stanton N.A. and Baber C. (2008) Modelling of human alarm handling response times: a case of the Ladbroke Grove rail accident in the UK. *Ergonomics*, 51:4, 423–40.

8
Human Factors Methods Integration: Operation Provide Comfort Friendly Fire Case Study

8.1 INTRODUCTION

The aim of this concluding analysis is to present an example of a comprehensive integrated framework of different Human Factors methods for evaluating accident scenarios. The case study presented demonstrates how the methods described in this book can be combined and integrated in order to provide a more exhaustive analysis of accidents occurring in complex sociotechnical systems. Applying the methods described in this book in isolation is perfectly acceptable for accident analysis purposes, and whilst their utility when applied alone is assured, Human Factors methods are increasingly being applied together as part of frameworks or integrated suites of methods for the evaluation of performance in complex sociotechnical systems (e.g. Stanton et al. 2005; Walker et al. 2006). Often accident scenarios are so complex and multi-faceted, and analysis requirements so diverse, that various methods need to be applied since one method in isolation cannot cater for the scenario and analysis requirements. The EAST framework (Stanton et al. 2005) described in Chapter 2 is one multi-method framework that was developed to examine the work of distributed teams in complex sociotechnical systems. The utility of the framework for system performance description and analysis is such that it has since been applied for varying purposes, including accident analysis.

8.1.1 Integrating Human Factors Methods

There are well over 100 Human Factors methods available covering all manner of concepts. In this chapter, an approach based on methods integration is proposed. The aim is to show how existing methods can be combined in useful ways to analyse complex, multi-faceted accident scenarios. Methods integration has a number of compelling advantages, because not only does the integration of existing methods bring reassurance in terms of a validation history, but it also enables the same data to be analysed from multiple perspectives. These multiple perspectives, as well as being inherent in the scenario that is being described, also provide a form of internal validity. Assuming that the separate methods integrate on a theoretical level, then their application to the same data set offers a form of 'analysis triangulation'.

It is well known that multiple failures of differing types, occurring over a period of time, are normally involved in the accidents that occur in complex sociotechnical systems (e.g. Hollnagel 2004; Rasmussen 1997; Reason 1990). Such failures might include communications failures, poor equipment design and maintenance, absence of appropriate procedures, and omissions and violations by front line workers, to name only a few. Developing a comprehensive understanding of the causal factors involved in accidents within complex sociotechnical systems is therefore likely to prove extremely difficult when only one method is applied. For example, using the methods described in this book as a reference point, applying CPA will prove useful for modelling appropriate operator response times; however, causal factors downstream from the accident are overlooked. AcciMap, on the other hand, will provide a neat overview of the failures across the entire organisational system; however, in-depth interrogation of the different failures involved is not supported.

More than likely then, none of the Human Factors methods described can, in isolation, exhaustively describe accident scenarios. Using an integrated suite of methods, however, allows scenarios to be analysed exhaustively from many perspectives. The EAST framework, e.g. takes a 'network of networks' approach to analysing activities taking place in complex sociotechnical systems. This allows activity to be analysed from three different but interlinked perspectives, the task, social and knowledge networks that underlie collaborative activity. Task networks represent a summary of the goals and subsequent tasks being performed within a system. Social networks analyse the organisation of the team and the communications taking place between the actors working in the team. Finally, knowledge networks describe the information and knowledge (distributed situation awareness) that the actors use and share in order to perform the teamwork activities in question. This 'network of networks' approach to understanding collaborative endeavour is represented in Figure 8-1 (adapted from Houghton et al. 2008).

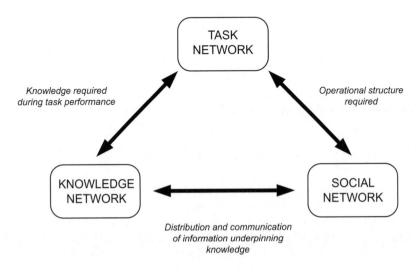

Figure 8-1   Network of networks approach

8.1.2 Summary of Component Methods

For the analysis presented in this chapter, we were interested in various aspects of performance prior to, and during, the accident scenario, including the specific tasks undertaken by those involved, communications between agents, situation awareness, teamwork, and the technology used to support communications. Accordingly an adapted version of the EAST approach described in Chapter 2 was used. The methods utilised included HTA (Annett et al. 1971), OSDs (Stanton et al. 2005), SNA (Driskell and Mullen 2004), the propositional network approach (Salmon et al. 2009), Communications Usage Diagram (CUD; Watts and Monk 2000) and Coordination Demands Analysis (CDA; Burke 2004). The framework used is represented in Figure 8-2.

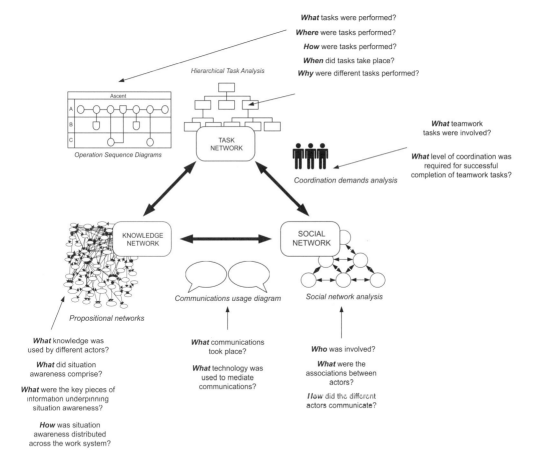

Figure 8-2 Network of network approach overlaid with methods applied during case study

This integrated approach allowed us to consider the following facets of the accident scenario analysed:

1. The physical and cognitive tasks involved (HTA, OSD);
2. Decision-making during the scenario (HTA, propositional networks);
3. Communications during the scenario (SNA, OSD, CUD);
4. The technology used to support communications (CUD);
5. Situation awareness during the scenario (propositional networks); and
6. The level of coordination between actors during the accident scenario (CDA).

The overlap between methods and the constructs they access is explained by the multiple perspectives provided on issues such as the 'who' and the 'what'. For example, the HTA deals with 'what' tasks and goals, the propositional networks deal with 'what' knowledge or situation awareness underpins the tasks being performed, and CDA deals with 'what' level of coordination was achieved by actors performing the tasks, each being a different but complementary perspective on the same data, which is an example of analysis triangulation.

8.2 INCIDENT DESCRIPTION

The incident in question occurred in Northern Iraq during Operation Provide Comfort (OPC), a multinational coalition operation designed to protect Kurdish refugees in the area from the Iraqi military. Numerous acronyms are utilised throughout in the remainder of this chapter. Table 8-1 below presents these acronyms along with their explanations.

Table 8-1   Table of acronyms

Acronym	Explanation of Acronym
ACE	Air Command Element
ACO	Air Control Order
ASO	Air Surveillance Officer
ATO	Air Tasking Order
AWACS	Airborne Warning and Control System
BH	Black Hawk – UH-60 Helicopter
CFAC	Combined Forces Air Component
CTF	Combined Task Force
En-route	En-route controller
F-15	US Fighter Jet
IFF	Identification Friend or Foe
JOIC	Joint Operations Intelligence Centre
MCC	Military Coordination Centre
OPC	Operation Provide Comfort
ROE	Rules of Engagement
SD	Senior Director
SPINS	Special Instructions
TAOR	Tactical Area of Responsibility
VID	Visual Identification

CHAPTER 8 • HUMAN FACTORS METHODS INTEGRATION: OPERATION PROVIDE COMFORT FRIENDLY FIRE CASE STUDY

A key component of OPC involved the construction and protection of a Tactical Area of Responsibility (TAOR), a secure area for the Kurdish refugees which included a No-Fly Zone (NFZ) prohibiting Iraqi aircraft from entering the area. The TAOR was enforced by a coalition Air Force and Army from a number of nations including America, Britain, Turkey and France. To maintain the security of the area, daily flight operations by Air Force jets were undertaken to ensure that the area was free from hostile aircraft and to protect it from any incursions. On 15 April 1994, during one such flight operation, two US Army Black Hawk helicopters were mistakenly identified as enemy 'Hind' helicopters and shot down by two friendly US F-15 fighter jets, killing all 26 personnel on board (USAF Accident Investigation Board 1994; United States General Accounting Office 1997).

Various agents were involved in the incident, both directly and indirectly, ranging from the F-15 and Black Hawk pilots to the Commander-in-Chief of Europe. A representation of the agents involved and the command and control structure adopted is presented in Figure 8-3.

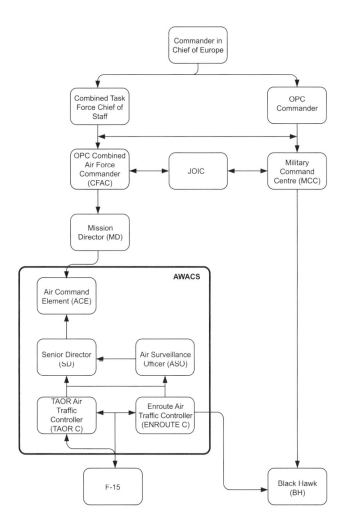

Figure 8-3   Command structure of the OPC (adapted from Leveson 2002)

Both sets of pilots (Black Hawk and F-15) were controlled by the AWACS team comprising a Mission Director, an Air Command Element, a Senior Director, an Air Surveillance Officer, TAOR Air Traffic Controller and an En-route Air Traffic Controller. The AWACS team was under the control of the OPC Combined Air Force Commander. In addition to being controlled by AWACS, the F-15 pilots were also controlled by CFAC, and the Black Hawk pilots were also controlled by the OPC Military Command Centre (MCC Army). Both the MCC and CFAC were controlled by two bodies – the Combined Task Force Chief of Staff and the OPC Commander. In turn, both these bodies were controlled by the Commander-in-Chief of Europe (USAF Accident Investigation Board 1994; United States General Accounting Office 1997).

In Table 8-2, a timeline of events for the incident is presented. The timeline was developed based on various accounts of the incident (e.g. Leveson 2002; Leveson et al. 2002; USAF Accident Investigation Board 1994; Snook 2000; United States General Accounting Office 1997).

Table 8-2   Timeline of events

Time	Event
09:21	Black Hawks reported entry into NFZ by radio to AWACS en-route controller.
09:21	The AWACS en-route controller received the Black Hawks' report and labelled them as friendly on the radar. The en-route controller did not tell the Black Hawks that they needed to change their IFF and use a different radio frequency now they were inside the NFZ.
09:24	Black Hawks landed just inside the NFZ and the radar and IFF signals diminished.
09:24	Due to the diminished returns the AWACS en-route controller removed the friendly labels from the radar.
09:36	F-15s took off from Incirlik.
09:36	The F-15s were checked by the en-route controller and the correct IFF code was verified.
09:54	The Black Hawks took off and informed the AWACS en-route controller of this. During this communication the Black Hawks were using a code system which the AWACS en-route controller did not understand.
09:54	After the radio communication from the Black Hawks the AWACS controller relabelled them as friendly on the radar.
10:05	The F-15s contacted AWACS en-route controller to inform him that they were just about to enter the NFZ.
10:11	The Black Hawks IFF and radar returns were again diminished, this time due to the large mountains they were flying through. This meant that the radar now displayed a friendly signal but without a Black Hawk identifying symbol.
10:13	The Air Surveillance Officer (ASO) noticed that the Black Hawks returns had diminished and sent a large arrow and flashing light message to the Senior Director based at the Black Hawks last location.
10:14	The Senior Director did not notice this message before it was automatically removed from the system.
10:15	At the last stage prior to entering the TAOR the F-15s contacted AWACS and were told that there were no updates on the situation, this time communicating with the Air Command Element (ACE).

Table 8-2   *Concluded*

10:20	Once they had entered the TAOR the F-15s again spoke to AWACS, this time using the TAOR radio frequency and thus communicating with the TAOR Air Traffic Controller.
10:20	One member of the AWACS team could see both the F-15s and Black Hawks but no communications were made regarding either aircraft's presence.
10:21	Believing that the Black Hawks had landed and upon agreement with the Senior Director the TAOR Air Traffic Controller (ATC) removed the friendly symbols that had been attached to the Black Hawks.
10:22	The F-15 pilots received unknown radar returns and reported these to the TAOR controller; the F-15 pilots were told that AWACS had no knowledge of radar returns in that area.
10:22	F-15 pilots used their IFF technology to attempt to identify the unknown radar returns. They received a negative response, followed by a short friendly response before returning to a negative response.
10:23	The Black Hawks' radar returns increased and were shown on the AWACS radar in the same place that the F-15s had reported unknown radar returns.
10:25	After receiving a second call from the F-15 pilots the AWACS stated that they could see radar returns in that area. No mention of the Black Hawks was made.
	The AWACS crew labelled the radar returns as unknown and attempted to identify the returns using their IFF technology.
10:28	The F-15s continued to attempt to identify the radar returns with both IFF interrogations and visual identification passes.
10.28	The visual identification pass, which was conducted at a speed far above that stated by the Rules of Engagement, resulted in the F-15 lead pilot identifying the radar returns as two Hind Helicopters. This was confirmed by the F-15 wing pilot on his visual identification pass. The AWACS crew were listening to the F-15 communications.
10:29	The F-15 lead reported his intent to engage the helicopters to AWACS.
10:30	The F-15 pilot conducted further IFF interrogation but received no friendly response and so he and the F-15 wing continued with the engagement and destroyed both helicopters.

The incident occurred because the F-15 pilots had no knowledge regarding the presence of the Black Hawk helicopters. The information about the Black Hawk flights was not included on the daily Air Tasking Order. The AWACS team did hold information regarding the presence of the Black Hawks but they did not advise the F-15 pilots of this. The Black Hawks were unable to hear communications between F-15 pilots and AWACS, or communicate with the F-15 pilots as they were using an incorrect radio frequency. As a result of these factors when the F-15 pilots identified radar contacts they assumed them to be enemy. An attempt to interrogate the contacts using Identification Friend or Foe (IFF) technology (a system that interrogates other assets to distinguish if they are friendly, unknown or enemy) failed because the Black Hawks were not informed of the correct IFF mode to use within the TAOR, further cementing the notion that the contacts were enemy. The F-15 pilots then conducted a visual identification but due to inadequate visual recognition training, and additional fuel tanks fitted to the Black Hawks changing their appearance, the Black Hawks were misidentified as Iraqi Hind Helicopters (USAF Accident Investigation Board 1994; United States General Accounting Office 1997).

The key causal factors outlined by the official government inquiry into the incident, along with previous analyses, are presented below:

1. The Combined Task Force failed to provide clear guidance leading to an unclear understanding of roles throughout the coalition force, specifically with regard to supporting helicopter missions;
2. The Combined Task Force did not view helicopters as part of the air operations and as such they were not integrated adequately or monitored sufficiently when flying in the TAOR;
3. There was an insufficient level of training given regarding the Rules of Engagement within the TAOR resulting in a simplified understanding of the ROE;
4. The AWACS team held information regarding the presence of the Black Hawks but did not pass this on to the F-15 pilots;
5. The Black Hawks were unaware of the correct IFF Mode or radio frequency to use inside the TAOR; and
6. Due to poor visual identification training, and additional fuel tanks on the Black Hawks, the F-15 pilots misidentified them as enemy.

8.3 DATA SOURCES AND DATA COLLECTION

Four main data sources were used to support the EAST analysis. These were the USAF Aircraft Accident Investigation Board report, the United States General Accounting Office Investigation report, and two other accounts of the incident (e.g. Leveson et al. 2002; Leveson 2002; Snook 2000).

8.4 ANALYSIS PROCEDURE AND RESOURCES INVESTED

The EAST framework was used to construct two analyses; the first was a description of the incident as it happened, whereas the second involved modelling an idealised version of events. The ideal scenario shows how events (e.g. information, communications, tasks) could have unfolded so that the friendly fire incident was avoided. The ideal scenario provides a benchmark against which to compare the incident. Rather than a search for failures, the comparison of the actual and ideal scenarios allows for the identification of all differences between the scenarios, with the differences identified representing possible causal factors. As Woods et al. (1994, 197) stated: 'the same factors govern the expression of both expertise and error'.

To identify causal factors both success and breakdowns must therefore be explored. In this case, the actual scenario represents the breakdowns and the ideal scenario represents the successes. The idealised version of the scenario was created by replacing all inappropriate tasks, communications and information elements that occurred within the actual scenario, with appropriate replacement actions, as well as inserting any missing actions. This was based upon extensive reading around the incident (Leveson 2002; Leveson et al. 2002; Piper 2001; Snook 2000; USAF Accident Investigation Board 1994; United States General Accounting Office 1997). For example, the USAF report clearly states actions, communications and beliefs that were faulty and also the roles, tasks and actions that should have occurred. This allowed the authors to systematically develop the 'ideal' scenario. To demonstrate, an example of the report explicitly stating an inappropriate action and the correct, appropriate action is given below.

Within the USAF report, the erroneous act of the Black Hawks operating with the incorrect IFF frequency for within the TAOR is described as follows:

> The F-15 pilots attempted to electronically identify the radar contacts by interrogating the ATO-designated IFF Mode I and Mode IV aircraft codes. The helicopter crew members were apparently not aware of the correct Mode I code specified for use within the TAOR and had the Mode I code specified for use outside the TAOR in their IFF transponders. The result was that the F-15s did not receive a Mode I response.

The USAF report references changes to the system that have now been introduced to prevent the incident happening again; these are actions that should have occurred but did not. With respect to the above example, the 'ideal' is referred to three times:

> All aircraft (including helicopters): Require contact with Airborne Warning and Control System (AWACS) on TAOR Ultra High Frequency (UHF). Have Quick or UHF clear radio frequencies and confirmation of Identification Friend or Foe (IFF) Modes I, II, and IV. If either negative radio contact with AWACS or inoperative Mode IV do not proceed into TAOR.

> Immediately after takeoff, contact AWACS and reconfirm IFF Modes I, II and IV are operating. All Aircraft (including helicopters): Must be under positive control (i.e. radio contact and positive IFF/SIF) of AWACS to operate inside the TAOR. Require positive IFF/Special Identification Feature (SIF) and radio checks be accomplished while enough fuel remains to return to Diyarbakir AB.

Within the report the IFF mode check is clearly stated as the role of AWACS weapon directors:

> One WD (weapons director) acts as an en-route controller, responsible for controlling the flow of aircraft to and from the TAOR. This person also conducts IFF and radio checks on all OPC aircraft.

8.5 OUTPUTS

8.5.1 Hierarchical Task Analysis (HTA)

HTA (Annett et al. 1971) describes the system under analysis in terms of a structured hierarchy of goals and sub-goals along with feedback loops (Annett 2004). One useful way of summarising large and complex HTA outputs is through the construction of a task network, which provides a summary of the main higher level goals and tasks involved. The task networks, one for the actual scenario as it unfolded, and one for the ideal scenario, are presented in Figure 8-4.

Within the actual scenario a linear chain of command was identified with the tasks highly separated and two separate agencies (CFAC and MCC) performing separate tasks.

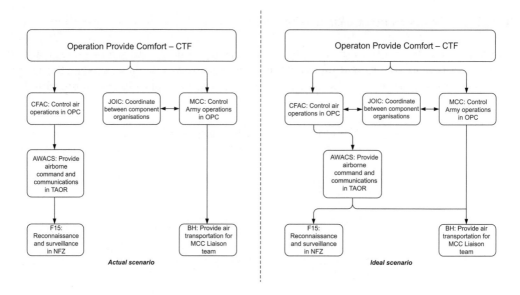

Figure 8-4 Task network for the actual and ideal scenarios

However, within the ideal scenario, the structure was far more networked with tasks being undertaken collaboratively. This represents both a higher level of collaboration and cooperation during the ideal scenario than during the actual scenario.

The HTAs also enable comparison of the two scenarios in terms of the activities that occurred. A number of core operations that should not or did not occur within the actual scenario were identified. This allowed for the development of a set of initiating factors (confirmed by the government report as well as a number of subsequent analyses of the same incident). In summary, there were 43 additional tasks that were not needed in the accident scenario and 66 tasks that should have, but did not, occur.

The ability to map these initiating factors onto previous analyses bodes well for the reliability of EAST and for its application to friendly fire incidents. An example of the cross-over between the non-compatible operations and the erroneous actions is operation 1.3: Compatible technology evolution, present in the ideal scenario HTA. This leads onto operation 1.3.1, 1.3.2 and 1.3.3 which ensure that all OPC aircraft are fitted with the same radio equipment and can communicate with one another ensuring that the Black Hawks and F-15s are able to use their radio equipment to communicate with one another. These operations are not present within the actual HTA and due to this the Black Hawks and F-15s were not able to communicate with one another – a key failure identified in the government inquiry report and other analyses of the incident (Leveson 2002; Leveson et al. 2002 Snook 2000).

Within this analysis the HTA is developed in order to identify initiating factors that may have led to the incident of fratricide, and throughout the remainder of the EAST analysis, the methods allow the analyst to delve deeper into these causal factors exploring the context surrounding them. Such exploration is important for understanding the origins of the initiating factors that led to the final act of misidentification. The HTA output also forms the basis for many of the other EAST analysis methods, such as coordination demands analysis and propositional networks.

CHAPTER 8 • HUMAN FACTORS METHODS INTEGRATION: OPERATION PROVIDE COMFORT FRIENDLY FIRE CASE STUDY

8.5.2 Social Network Analysis (SNA)

Social network diagrams were constructed for the actual and ideal scenarios. The actual scenario social network, depicting the communications that took place between the agents involved, is presented in Figure 8-5. The ideal scenario social network is presented in Figure 8-6.

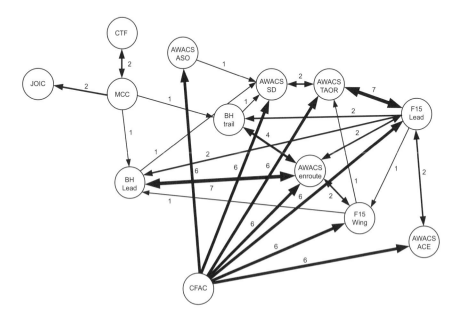

Figure 8-5 Actual scenario social network

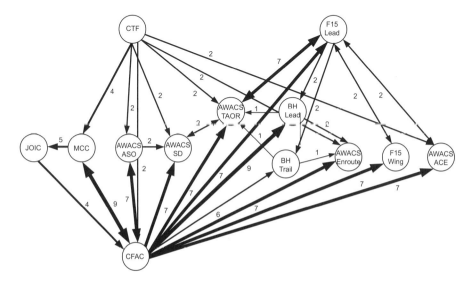

Figure 8-6 Ideal scenario social network

Examination of the social networks reveals key differences between the social organisation and communications taking place between agents in the actual and ideal scenarios. In the ideal scenario three key agents can be identified – CTF, CFAC and F-15 lead. These three agents possess communications links with most of the other agents involved. The actual accident scenario, however, appears more convoluted with many agents having links to many other agents and a lower level of key, or central, agents with fewer, stronger links. From this we can surmise that the ideal social network is more centrally organised around a number of key agents – utilising only a few strong communication links – whereas the actual scenario network is more distributed with a greater number of weaker communication links.

In addition to providing a useful graphical representation of the communications between agents, social networks can be analysed mathematically in order to interrogate them for aspects such as density, key agents and paths through the networks. In this case the following metrics were used to interrogate the networks: sociometric status, centrality, density and cohesion. A description of each metric, along with the analysis outputs, is presented below.

Network Density Network density represents the level of interconnectivity of the network in terms of communications links between agents. The formula for network density is below (adapted from Walker et al. 2011).

$$\text{Network Density} = \frac{2e}{n(n-1)}$$

Where:
e = number of links in network
n = number of information elements in network

Formula 8-1   Network density

Network density is expressed as a value between zero and one, with zero representing a network with no connections between agents, and one representing a network in which every agent is connected to every other agent (Kakimoto et al. 2006, cited in Walker et al. 2011). Denser networks are therefore characteristic of more communications links between agents. In this case the analysis shows that both networks have similar levels of density, with the actual scenario network producing a density score of 0.21, and the ideal scenario 0.23. Both values are indicative of moderate levels of connectivity within the networks.

Sociometric Status Sociometric status provides a measure of how 'busy' a node is relative to the total number of nodes present within the network under analysis (Houghton et al. 2006). Sociometric status is calculated using the following formula (g is the total number of nodes in the network, i and j are individual nodes and are the edge values from node i to node j).

$$\text{Status} = \frac{1}{g-1} \sum_{j=1}^{g} \left(x_{ji} + x_{ij} \right)$$

Formula 8-2 Sociometric status formula

In practical terms then, sociometric status gives an indication of the relative prominence an individual agent has as a communicator with others in the network (Houghton et al. 2006). Agents with high sociometric values are highly connected to others within the network, whereas nodes with low sociometric status values are likely to reside on the periphery of the network and have low connectedness with others in the network.

The sociometric status analysis presented in Figure 8-7 indicated that the key agent in both actual and ideal scenarios (i.e. with the highest sociometric status) is the CFAC; however, the level of sociometric status differs greatly between the two scenarios. This difference is clearly illustrated in Figure 8-7, which shows that in the actual scenario CFAC has a value of 3.5 compared to 6.7 in the ideal scenario. From this we can surmise that for the scenario to run smoothly and not lead to a fratricide incident, the CFAC must take a more prominent communications role within the system.

The idea that agents working at the higher levels of the overall system, such as CFAC, should take a more prominent role in the scenario in order to prevent fratricide is further illustrated by the sociometric status metrics of other agents working at the upper levels of the system. Within the actual scenario both MCC and CTF have low sociometric status values (0.5 and 0.16 respectively) whereas in the ideal scenario the sociometric status values of these higher level agents are higher (1.5 and 1.3 respectively). The sociometric status values appear to highlight the need for the three higher level organisations (CTF, CFAC and MCC) to play a more prominent role in the scenario in order to prevent the fratricide incident.

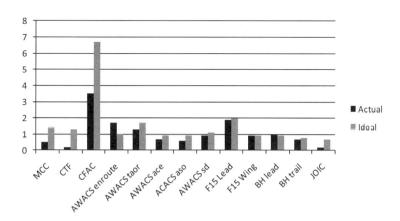

Figure 8-7 Sociometric status for actual and ideal scenarios

The sociometric status for the actual and ideal also enables comparison of the mean sociometric status value within the networks. The mean sociometric status for the actual scenario is 1.0897, whereas for the ideal scenario it is higher at 1.5769. In addition, the ideal scenario has a higher sociometric status for all actors except the en-route controller and the Black Hawk lead pilot. This suggests that the ideal scenario was characterised by a high level of communications contributions for the majority of actors. The lower level of sociometric status found for the en-route controller and Black Hawk lead pilot are representative of the increased role these agents played in the actual scenario due to the en-route controller maintaining control of the Black Hawk helicopters.

Centrality Centrality is also a metric of the standing of a node within a network (Houghton et al. 2006), but here this standing is in terms of its 'distance' from all other nodes in the network. A central node is one that is close to all other nodes in the network and a message conveyed from that node to an arbitrarily selected other node in the network would, on average, arrive via the least number of relaying hops (Houghton et al. 2006). The formula for centrality is presented below.

$$Centrality = \frac{\sum_{i=1; j=1}^{g} \delta ij}{\sum_{j=1}^{g} (\delta ij + \delta ji)}$$

Formula 8-3   Centrality formula

The higher the centrality value for an agent the greater the ease of information flow within the system. The centrality analysis outputs (see Figure 8-8) reflect the sociometric status outputs, with both the CTF and CFAC having higher centrality rates in the idealised scenario (7.4 and 10.5 respectively) than in the actual scenario (4.5 and 7.1 respectively). The actual scenario is characterised by higher centrality figures for the lower systemic levels with the highest four values assigned to the F-15 and Black Hawk pilots. In contrast to this, the highest four values in the idealised scenario are CFAC, CTF, AWACS TAOR controller and the F-15 lead.

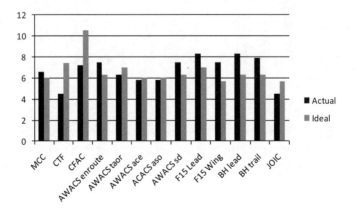

Figure 8-8   Centrality values for actual and ideal scenarios

The SNA highlights key communication differences between the two systems. Similar levels of density were found for both networks suggesting that both scenarios contained similar levels of communication connections. Importantly, the other metrics used highlighted that it was the way in which these connections were utilised that differed between the two scenarios. The sociometric status results suggest that within the ideal scenario a greater number of agents made a greater contribution in terms of communications compared to the actual scenario, which results in information being able to pass through the network more easily. From this we can conclude that higher communication levels may act as a preventative measure in fratricide incidents, and conversely that low levels of communication within a system may be a key causal factor involved in fratricide incidents. Specifically, lower levels of contribution were found at the higher organisational levels (such as CFAC and CTF) of the system. This tentatively suggests that low levels of communication from higher organisational levels may be a causal factor in fratricide incidents. The centrality results provide further evidence to support this assertion with higher centrality values for agents at the higher levels of the organisational system within the ideal scenario compared to the actual scenario, suggesting that within the ideal scenario agents at higher levels were more connected to other agents in the system.

8.5.3 Communications Usage Diagram (CUD)

CUD (Watts and Monk 2000) is used to represent communications and the technology used to mediate communications. In this case, bottom level task steps from the HTA description of the scenarios involving communications between agents are described in terms of the agents involved, their role in the overall work system, and the communication media through which the communication act occurred (e.g. radio, face-to-face, paper). An extract of the CUD analysis is presented in Table 8-3.

In both scenarios three types of communications media were utilised: radio communications, face to face verbal communications, and paper based communications, such as distribution of mission orders.

Figure 8-9 below provides a representation of the frequency with which each communications media was used during the actual and ideal scenario.

Radio communications were the most commonly used form of communication in both the actual and ideal scenarios. The greatest difference between scenarios is in the paper communication modality suggesting that an increase in the distribution of orders etc. would have been beneficial in the actual scenario. The CUD analysis also allows comparison of overall communication levels between the two scenarios. The analysis identified higher levels of communications within the ideal scenario, compared to the actual scenario, suggesting that low levels of communication may be a causal factor in fratricide incidents.

8.5.4 Coordination Demands Analysis (CDA)

CDA (Burke 2004) provides a quantitative rating of the level of coordination exhibited between team members during collaborative scenarios. Again using the HTA as its primary input, coordination between team members on each collaborative bottom level task step was rated on a number of key dimensions (see Table 8-4).

Table 8-3   Actual Scenario CUD analysis extract

Stage	Action	Agent	Work position 1	Work position 2	Agent	Radio	Face to Face	Paper
1.2.1 and 3.1.1	Receive request for BH mission	MCC	MCC	CTF	CTF	1		
1.2.2 and 3.1.2	Approve request for BH mission	CTF	CTF	MCC	MCC	1		
2.1.1	ROE Training	CFAC	CFAC				1	
2.1.2.1	ROE briefing	CFAC	CFAC				1	
2.1.2.2.2	Publish SPINS brief	CFAC	CFAC					1
2.1.2.3.2	Publish ACO brief	CFAC	CFAC					1
2.1.2.4.2	Publish ATO brief	CFAC	CFAC					1
2.2.1	CFAC scheduling meeting	CFAC	CFAC				1	
2.2.2.2	Publish CFAC updates	CFAC	CFAC					1
3.1.3.1	ROE brief	MCC	MCC	BH	BH		1	
3.1.3.2.2	Publish ATO of army ops	MCC	MCC	BH	BH			1
3.1.3.3.2	Publish ACO of army ops	MCC	MCC	BH	BH			1
3.1.3.4.2	Publish SPINS of army ops	MCC	MCC	BH	BH			1
3.2.1.2 and 7.1	Send MCC schedule to JOIC	MCC	MCC	JOIC	JOIC			1
3.2.2.2 and 7.2	Send MCC SITREP to JOIC	MCC	MCC	JOIC	JOIC			1

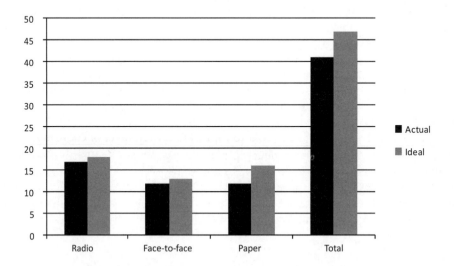

Figure 8-9   Usage of different communications media

Table 8-4   CDA teamwork taxonomy (adapted from Burke 2004)

Coordination Dimension	Definition
Communication	Includes sending, receiving, and acknowledging information among crewmembers.
Situational Awareness (SA)	Refers to identifying the source and nature of problems, maintaining an accurate perception of the external environment, and detecting situations that require action.
Decision-making (DM)	Includes identifying possible solutions to problems, evaluating the consequences of each alternative, selecting the best alternative, and gathering information needed prior to arriving at a decision.
Mission analysis (MA)	Includes monitoring, allocating, and coordinating the resources of the team; prioritising tasks; setting goals and developing plans to accomplish the goals; creating contingency plans.
Leadership	Refers to directing activities of others, monitoring and assessing the performance of team members, motivating members, and communicating mission requirements.
Adaptability	Refers to the ability to alter one's course of action as necessary, maintain constructive behaviour under pressure, and adapt to internal or external changes.
Assertiveness	Refers to the willingness to make decisions, demonstrating initiative, and maintaining one's position until convinced otherwise by facts.
Total Coordination	Refers to the overall need for interaction and coordination among crew members.

Initially, tasks were classified as teamwork tasks (i.e. involving collaboration between two or more team members) or taskwork tasks (i.e. tasks completed in isolation by individuals only). For the actual scenario, 61 per cent of the task steps involved teamwork, with the remaining 39 per cent being individual or task work tasks. Similarly, in the ideal scenario, 65 per cent of the task steps involved teamwork, with the remaining 35 per cent involving task work only. This suggests that there is a comparable level of teamwork tasks in both scenarios.

Following the task step classification the level of coordination exhibited on each teamwork task step was rated on a scale of 1–3 for the eight dimensions presented in Table 8-4 (adapted from Burke 2004).

A total coordination score is then derived from the mean of the component scores. A rating of 2.25 (75 per cent) is deemed to represent a high level of coordinated activity (Stanton, Baber and Harris, 2008). The CDA ratings for the actual and ideal scenarios are presented in Figure 8-10.

During the ideal scenario, there were higher levels of coordination within teamwork tasks. From this it is concluded that although there is not a greater amount of teamwork tasks requiring coordination in the ideal scenario, there is a greater level of coordination required between team members whilst conducting the teamwork tasks.

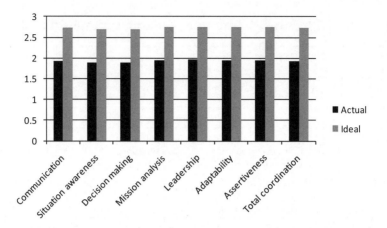

Figure 8-10 Coordination demands analysis coordination dimensions

8.5.5 Propositional Networks

To construct propositional networks, a content analysis was performed on the documentation surrounding the incident. The purpose of this was to identify key concepts or information elements and the relationships between them throughout the scenario. This allows a representation of the systems situation awareness throughout each scenario to be developed, following which the ownership, usage and communication of the information elements by each agent can be interrogated.

Propositional networks were developed for both scenarios, including an overall propositional network representing situation awareness during the entire scenario and propositional networks covering 11 distinct scenario phases. Each incident phase network was colour coded to represent the usage of information elements by the different agents involved. This allowed the usage of information to be traced through the scenario on a temporal basis. For example, the propositional networks representing situation awareness at the onset of both scenarios (i.e. at 8.22) are presented in Figures 8-11 and 8-12.

The information elements have been roughly organised into sections dependent upon the agents that actively know the information. This allows one to clearly identify that within the actual scenario the F-15 pilots have no knowledge of the Black Hawk mission, except that there may be a Black Hawk flight at some point on that date. The F-15 pilots in the actual scenario also have knowledge that tells them that no flight may enter the TAOR before their sweep, and that all flights must be tracked by AWACS.

Further interrogation of the propositional networks is achieved through the use of network analysis statistics. Sociometric status, e.g. can be used to identify the key information elements underpinning situation awareness (Salmon et al. 2009). The key information elements were identified in this case using sociometric status calculations. The key information elements (Table 8-5) represent those with a value higher than the mean plus one standard deviation.

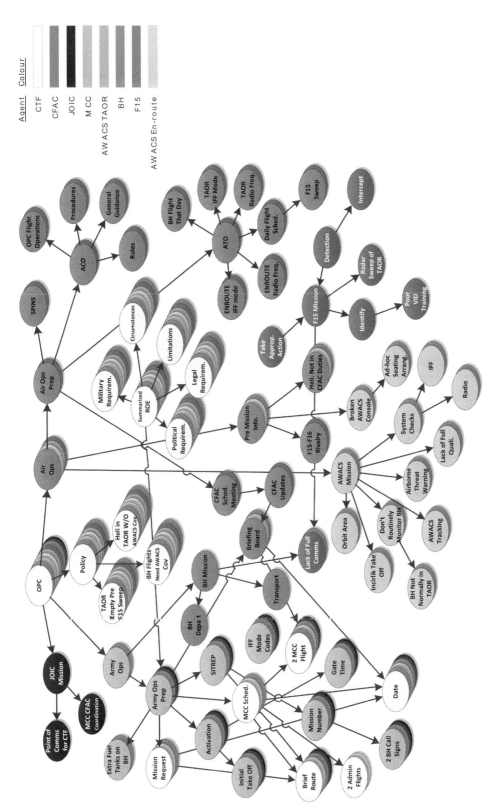

Figure 8-12   Propositional network representing systems situation awareness at 8.22 in the ideal scenario

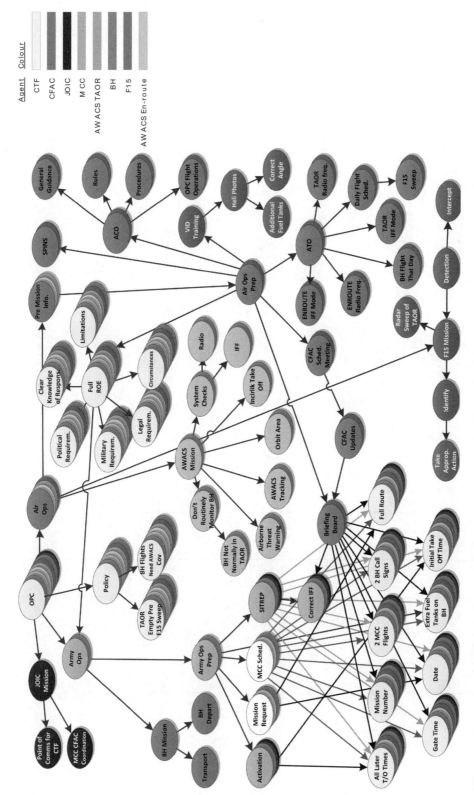

Figure 8-11 Propositional network representing systems situation awareness at 8.22 in the actual scenario

Table 8-5   Key information elements for both the actual and ideal scenarios

Key Information Element	Actual	Ideal
Summarised ROE	■	
Army Ops Prep	■	■
AWACS mission	■	■
ATO	■	■
Air Ops Prep	■	■
F-15 mission	■	■
Air Ops	■	
ACO	■	
Army Ops	■	
SITREP	■	■
BH mission	■	
EN-ROUTE IFF mode	■	
EN-ROUTE radio freq.	■	
AWACS tracking	■	■
F-15 report contact	■	
Briefing board		■
MCC schedule		■
Full ROE		■

Both the actual and ideal scenarios share seven key information elements: Army ops prep, AWACS mission, ATO, Air ops prep, F-15 mission, SITREP and AWACS tracking. The ideal scenario contains three additional key information elements: Briefing board, MCC schedule and full ROE. The briefing board in the F-15 briefing room contained information about the Black Hawk flights; within the actual scenario this information was not disseminated to the F-15 pilots. In the ideal scenario the F-15 pilots did receive this information and thus were prepared to see friendly aircraft within the TAOR, this was a key information element within the scenario and also a key piece of information in preventing the shoot-down of the Black Hawks. The MCC schedule also contained information regarding the Black Hawk mission; if this information had been disseminated properly, as in the ideal scenario, then all agents would have been aware of the Black Hawk flight, again preventing the Black Hawk shoot-down.

The most significant difference between the two scenarios is the information element of the full ROE (in the ideal) scenario compared to the summarised ROE (in the actual scenario). The summarised ROE meant that the F-15 pilots and AWACS staff had an inaccurate understanding of the Rules of Engagement for the situation they were in. This led to the F-15 pilots shooting down the Black Hawk helicopters due to a lack of guidance from AWACS. In the ideal scenario, in which all participants had an understanding of the full ROE, this would not have occurred. The F-15 pilots would have known that they were not allowed to fire, and they also would have received guidance on what to do from the AWACS staff.

The actual scenario contained an additional eight information elements, including the summarised ROE discussed above, as well as Air ops, ACO, Army Ops, BH mission, En-route IFF mode, En-route radio frequency and F-15 report contact. These additional information elements were not needed as key information elements within the ideal scenario, e.g. the actual scenario contained en-route IFF mode and en-route radio frequency, which were not key in the ideal scenario as the Black Hawk pilots changed to the correct TAOR IFF mode and radio frequency once entering the TAOR.

In order to explore the development of situation awareness throughout each scenario, they were broken down into eleven distinct phases based on the key events that unfolded. Propositional networks were developed for each phase, for both the actual and ideal scenarios. A summary of the key information elements during each phase is presented in Tables 8-6 (actual scenario) and 8-7 (ideal scenario).

Table 8-6   Actual scenario key information elements per phase

Information element	Phase										
	Phase 1 (08:22)	Phase 2 (09:21)	Phase 3 (09:27)	Phase 4 (09:54)	Phase 5 (10.12)	Phase 6 (10:15)	Phase 7 (10:20)	Phase 8 (10:21)	Phase 9 (10:22)	Phase 10 (10:24)	Phase 11 (09:28)
Summarised ROE	X	X	X	X	X	X	X	X	X	X	X
Army Ops Prep	X	X	X	X	X	X	X	X	X	X	X
AWACS mission	X	X	X	X	X	X	X	X	X	X	X
ATO	X	X	X	X	X	X	X	X	X	X	X
Air Ops Prep	X	X	X	X	X	X	X	X	X	X	X
F-15 mission	X	X	X	X	X	X	X	X	X	X	X
Air Ops	X	X	X	X	X	X	X	X	X	X	X
ACO	X	X	X	X	X	X	X	X	X	X	X
Army Ops	X	X	X	X	X	X	X	X	X	X	X
SITREP	X	X	X	X	X	X	X	X	X	X	X
Black Hawk mission		X	X	X	X	X	X	X	X	X	X
En route IFF mode		X			X						
En route radio frequency		X			X						
AWACS tracking		X							X		X
F-15 contact report									X		

CHAPTER 8 • HUMAN FACTORS METHODS INTEGRATION: OPERATION PROVIDE COMFORT FRIENDLY FIRE CASE STUDY

Table 8-7   Ideal scenario key information elements per scenario phase

Information element	Phase 1 (08:22)	Phase 2 (09:21)	Phase 3 (09:27)	Phase 4 (09:54)	Phase 5 (10.12)	Phase 6 (10:15)	Phase 7 (10:20)	Phase 8 (10:21)	Phase 9 (10:22)	Phase 10 (10:24)	Phase 11 (09:28)
Full ROE	✓	✓	✓	✓	✓	✓	✓	✓	✓	✓	✓
Army Ops Prep	✓	✓	✓	✓	✓	✓	✓	✓	✓	✓	✓
AWACS mission	✓	✓	✓	✓	✓	✓	✓	✓	✓	✓	✓
ATO	✓	✓	✓	✓	✓	✓	✓	✓	✓	✓	✓
Air Ops Prep	✓	✓	✓	✓	✓	✓	✓	✓	✓	✓	✓
F-15 mission	✓						✓	✓	✓	✓	✓
SITREP	✓	✓	✓	✓	✓	✓	✓	✓	✓	✓	✓
Briefing board	✓	✓	✓	✓	✓	✓	✓	✓	✓	✓	✓
MCC schedule	✓	✓	✓	✓	✓	✓	✓	✓	✓	✓	✓
AWACS tracking	✓		✓		✓		✓	✓	✓	✓	✓

Table 8-6 illustrates the development of situational awareness throughout the scenario. The majority of the key information elements remained stable throughout the scenario; however, the information elements AWACS tracking, En-route IFF mode and En-route radio frequency were only key information elements during the phases of the scenario in which AWACS were actively tracking the Black Hawk helicopters, who were using the En-route IFF and radio frequency. The table also highlights the lack of BH mission as a key information element before the Black Hawks entered the TAOR at 09:21. This led to the assumption that knowledge of the Black Hawk mission was not disseminated throughout the system before the Black Hawks entered the TAOR. F-15 report contact became a key information element as soon as the contact was reported; this is as would be expected.

Table 8-7 illustrates the development of situational awareness within the ideal scenario by presenting the key information elements active each scenario phase.

Table 8-7 demonstrates that the majority of key information elements remained stable throughout the ideal scenario. F-15 mission only became a key information element after the F-15s took off at 10.15. AWACS's tracking is present more prominently within the ideal scenario than it was within the actual scenario.

8.5.6 Operation Sequence Diagrams (OSDs)

OSDs are used in this case to present a graphical illustration of the temporal sequence of events that occurred, in terms of who was involved and at which point in time different events occurred throughout the scenario. This allows for a simple, systematic interpretation of the scenario including the temporal structure and interrelations between processes. In this manner, the OSD provides a summarised representation of the HTA, CDA, CUD and SNA outputs, providing an illustration of the tasks, actors, communications, social organisation and the sequential and temporal elements of the scenario. Due to size restrictions, the full OSDs are not presented here; however, summaries of the OSDs are presented in the form of operational loading calculations, which represent the frequency with which the different agents were involved in different task types (e.g. operation, transmit, receive, request). The operational loading figures for the actual scenario are presented in Table 8-8, followed by ideal scenario operational loading figures in Table 8-9.

As illustrated in Tables 8-8 and 8-9, the total number of operations is higher in the ideal scenario than in the actual scenario. This suggests that there were higher levels of workload within the ideal scenario. One would expect higher levels of workload in the actual scenario as at times of high workload as research has demonstrated that, at times of high workload, teams are the greatest risk of making errors (Salas et al. 2005). However, closer examination of the data shows that the ideal scenario places greater levels of workload on agents at the higher levels of the system, including the MCC, CTF and especially CFAC, whilst lowering the levels of individual actors lower down the systemic levels such as the en-route controller and the Black Hawk trail pilot. It could be suggested that this provides a more equal distribution of workload across the system.

Table 8-8   Operation loading table for actual scenario

	OPERATION	TRANSMIT	RECEIVE	REQUEST	TOTAL
MCC	2	10	2	0	14
CTF	0	2	2	0	4
CFAC	4	84	0	0	88
EN-ROUTE	10	8	28	0	46
TAOR	6	8	24	1	39
ACE	5	2	14	0	21
ASO	5	1	12	0	18
SD	6	2	19	0	27
F-15 LEAD	8	23	22	1	54
F-15 WING	6	5	16	0	27
BH LEAD	7	10	9	0	26
BH TRAIL	3	6	9	0	18
JOIC	0	0	4	0	4
TOTAL	62	161	161	2	386

Table 8-9 Operation loading table for ideal scenario

	OPERATION	TRANSMIT	RECEIVE	TOTAL
MCC	4	14	17	35
CTF	0	30	2	32
CFAC	15	121	14	150
EN-ROUTE	7	0	22	29
TAOR	8	11	25	44
SD	5	1	19	25
ACE	5	1	17	23
ASO	5	2	16	23
F-15 LEAD	10	13	26	49
F-15 WING	7	3	17	27
BH LEAD	3	8	14	25
BH TRAIL	3	7	14	24
JOIC	0	6	10	16
TOTAL	72	217	213	502

8.6 DISCUSSION

Through the HTA a number of core operations that should not or did not occur within the accident scenario were identified. This allowed for the development of a set of initiating factors (confirmed by the government report as well as a number of academic analyses of the same incident). Through the CDA, coordinated activity was identified as critical for the prevention of fratricide, with the actual scenario showing lower levels of coordination than the ideal scenario. The SNA outputs demonstrated the need for agents at the higher levels of the organisation (such as CFAC and CTF) to contribute more communications to those at the lower system levels (F-15s and Black Hawks) in order to prevent an incident of fratricide. Propositional networks revealed the way in which actors within the scenario used the information available to them, along with the key information elements underpinning situation awareness throughout the scenarios. Specifically it was revealed that a number of key information elements were not available in the actual scenario but were available in the ideal scenario. For example, distribution of key information elements such as the MCC SITREP would potentially have prevented the incident from occurring. The presence, or availability, of information within a system does not guarantee that the information will be processed. Activation of knowledge is very important and allows us to illustrate what people were aware of, and what information they require (Stanton et al. 2006; Walker et al. 2006). EAST is able to not only illustrate the information within a system, but through its information networks it is able to identify which information elements are activated, by whom, and when.

8.7 SUMMARY OF KEY FINDINGS

The aim of this chapter was to demonstrate the use of the EAST framework for exhaustively examining the causal factors involved in the accidents that occur in complex sociotechnical systems. EAST was applied to both the actual accident scenario and to an ideal version of the scenario in which the incident of fratricide could not have occurred. The comparison of results enabled key insights into the differences between effective and defective performance and subsequently into the possible causality behind incidents of fratricide. Through the use of the methods that EAST incorporates, numerous analytical perspectives were applied to the fratricide incident resulting in a rich understanding of the situation including the social, information and task networks involved.

EAST was able to provide not only a discussion of the core factors involved in the incident but also quantitative measurements of these factors. This enables statistical comparison of the causal factors between teams and between scenarios enabling a greater depth of exploration of the causal factors involved in accidents. Such quantitative values can also be used in future analyses to ascertain any similar measures across incidents. Furthermore, the analysis allows for an exploration of the manner in which these causal factors interact with one another.

9
Discussion

9.1 INTRODUCTION

The purpose of this final chapter is to compare the accident analysis methods presented in this book in terms of their outputs when applied for accident analysis purposes. As discussed earlier, each method presented is unique in many ways, e.g. in terms of theoretical underpinning, the 'type' of accident analysis method it represents, the procedure applied, and the outputs derived. Aside from pertinent questions already considered surrounding issues such as the resources required to apply a particular method, the level of training required for analysts, and the reliability and validity of the method's outputs, the one key remaining question is, which of the methods described is the most useful when used for accident analysis purposes?

9.2 COMPARISON OF METHODS

Of course, this question has many dimensions, and the answer is dependent on a range of factors. For example, the utility of each approach depends heavily on the aims of the analysis and the resources available (e.g. time, analysts). For example, for those wishing to describe all of the causal factors involved in a particular accident across the entire sociotechnical system in which it occurred, a systems approach such as STAMP or AcciMap is logically the most efficient; however, when focusing on specific actions and a particular human operator's performance in a particular accident scenario, more task and operator focused approaches such as CPA or CDM might be appropriate. Further, if limited resources are available for the analysis, AcciMap is likely to more efficient than STAMP, which requires more resources in terms of time invested and analyst training.

We will therefore attempt to answer the question of which method is the most useful in the simplest way possible: that is, when applied to the same accident with infinite resources available, which method will produce the most exhaustive output in relation to modern day models of accident causation?

The answer to this question can be summarised neatly using Rasmussen's risk management framework described earlier. Figure 9-1 shows the likely outputs derived from each method mapped onto Rasmussen's risk management framework with regard to the extent to which they cover each organisational level described. In Figure 9-1, the bars and corresponding level on Rasmussen's model reflect a capability for the method in question to identify, given infinite resources, accident causation factors up to and including that particular level. For example, a systems approach such as AcciMap has the capability to identify failures across all of the levels specified by Rasmussen, whereas an operator-focused approach such as CPA focuses only on single operator performance on a specific series of work tasks. Holes in the bars, represented by the white circles, are reflective of likely omissions (i.e. a failure of the method to comprehensively cover all of the failures at that level) when analysts operate strictly within the confines of the method in question. For example, since HFACS provides finite taxonomies of failure modes, it is likely that, in domains other than aviation, analysts will be restricted in the contributing factors that they can identify (i.e. some failures will exist outside of the taxonomy).

As shown in Figure 9 1, the methods described are very different. Starting from the left of the figure, Accimap and STAMP are both systems approaches in that they attempt to identify failures and causal factors across the entire complex sociotechnical system. As such, given appropriate data they can potentially cover all of the levels specified by Rasmussen's model. Although FTA is not typically labelled a systems approach, its flexibility is such that it too can conceivably identify and represent failures across all six levels of the system. In practice, however, FTA is more often than not limited to failures on the front line and the human and hardware failures involved. The CDM approach typically focuses on decision making processes and the factors influencing them and so is capable of identifying failures at all levels, at least from the point of view of the person subject to interview. The analysis is therefore likely to be comprehensive in terms of failures at the work and staff levels; however, for the other levels it is likely to be less comprehensive since it reflects only the opinion and knowledge of the individual operator and does not typically consider individuals at other levels of the system.

Moving along the figure, HFACS covers the failures at the four levels up to and including the organisational influences level, and so does not typically consider failures at the governmental and local authority levels (although recent versions of the method have been modified to do so, e.g. Rashid et al. 2010). Further, since finite taxonomies are used, failures outside of the scope of these taxonomies can often be found at all four levels. EAST and TRACEr are the next methods along and both are grouped as covering failures up to and including the company/organisational level in Rasmussen's model. Although EAST uses a suite of different methods, it is often difficult to identify and represent failures at the higher governmental levels since the method is more concerned with the task, social and knowledge networks involved during collaborative work performance on the front line. Similarly, TRACEr focuses specifically on the errors made by individual operators on the front line, but does include performance shaping factors up to and including the company level.

SNA and propositional networks are next, covering failures up to and including the management level in Rasmussen's model. SNA typically focuses on the communications between workers, supervisors and managers during task performance and often does not include communications preceding the scenario under analysis (although this could be achieved given appropriate data; however, it is most often difficult to obtain data on the communications occurring at the higher organisational levels). Propositional networks are concerned with situation awareness only, and so cover the awareness held by the system before and during the accident scenario. The networks are therefore likely to be comprehensive for the levels involved in the activity under analysis (e.g. staff, work, management); however, it is highly difficult when using this approach to identify situation awareness failures at the higher company, regulatory and governmental levels, since the analysis is most often concentrated on the events immediately prior to and during the accident, not on situation awareness in the weeks and months preceding the accident (when situation awareness failures at the higher organisational levels are most likely to occur). Finally, CPA is said to cover the work level only; it typically focuses on the tasks undertaken by front line operators, specifically their ordering and the projected performance time for different task sequences.

The summary presented therefore shows the importance of clearly defining the analysis aims and the data available before selecting an appropriate accident analysis method. Only then can the most appropriate accident analysis method be selected with any surety. For practitioners working in the area of accident analysis, this suggests that experience and training in the application of various accident analysis methods, covering the spectrum of concepts likely to require investigation during accident analysis efforts (e.g. error, decision-making, situation awareness), is pertinent.

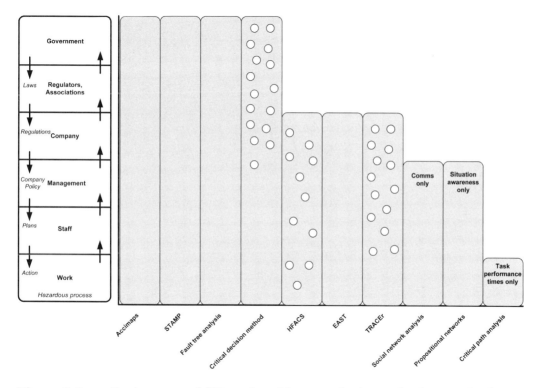

Figure 9-1   Systems capability of accident analysis methods; graph shows extent to which each method considers contributory factors at each of the levels described in Rasmussen's framework. White holes denote instances where the method in question is likely to be unable to fully cover all of the contributory factors at that level (e.g. by providing analysts with a limited taxonomy of failure modes)

References

Alper, S.J. and Karsh, B.T. (2009) A systematic review of safety violations in industry. *Accident Analysis and Prevention*, 41, 739–54.

Annett, J. (2004) Hierarchical task analysis. In: D. Diaper and N. Stanton (eds), *The Handbook of Task Analysis for Human Computer Interaction*. Mahwah, NJ: Lawrence Erlbaum Associates, pp. 67–82.

Annett, J. and Stanton, N.A. (2000) *Task Analysis*. London, UK: Taylor and Francis.

Annett, J., Duncan, K.D., Stammers, R.B. and Gray, M. (1971) *Task Analysis*. London: HMSO.

Arnold, R. (2009) A qualitative comparative analysis of SOAM and STAMP in ATM occurrence investigation. Unpublished MSc thesis, <http://sunnyday.mit.edu/Arnold-Thesis.pdf>, accessed 27 Jan 2011.

Baber, C. (2004) Critical path analysis for multimodal activity. In: N. Stanton, A. Hedge, K. Brookhuis, E. Salas and H. Hendrick (eds), *Handbook of Human Factors and Ergonomics Methods*. Boca Raton, FL: CRC Press, pp. 41.1–41.8.

Baber, C. and Mellor, B. (2001) Using critical path analysis to model multimodal human-computer interaction. *International Journal of Human-computer Studies*, 54, 613–36.

Baysari, M.T., Caponecchia, C., McIntosh, A.S. and Wilson, J.R. (2009) Classification of errors contributing to rail incidents and accidents: a comparison of two human error identification techniques. *Safety Science*, 47:7, 948–57.

Baysari, M.T., McIntosh, A.S. and Wilson, J. (2008) Understanding the human factors contribution to railway accidents and incidents in australia. *Accident Analysis and Prevention*, 40, 1750–7.

BBC News (2011) UK military deaths in Afghanistan and Iraq. <http://www.bbc.co.uk/news/uk-10629358>, accessed 14 January 2011.

Bennett, J.D., Passmore, D.L. (1985) Multinomial logit analysis of injury severity in US underground bituminous coal mines, 1975–1982. *Accident Analysis and Prevention*, 17:5, 399–408.

Billiton, B.H.P. (2005) ICAM investigation guideline. Issue 3, guideline number G44.

Blandford, A. and Wong, B.L.W. (2004) Situation awareness in emergency medical dispatch. *International Journal of Human-Computer Studies*, 61:1, 421–52.

Bomel Consortium (2003) The factors and causes contributing to fatal accidents 1996/97 to 2000/01. Summary report. HSE Task ID BOM\0040. C998\01\117R, Rev B, November 2003.

Brazier, A. and Ward, R. (2004) Different types of supervision and the impact on safety in the chemical and allied industries: assessment methodology and user guide. Health and safety executive report.

Brazier, A., Gait, A. and Waite, P. (2004) Different types of supervision and the impact on safety in the chemical and allied industries. Health and safety executive report.

Brookes, A., Smith, M. and Corkill, B. (2009) Report to the Trustees of the Sir Edmund Hillary Outdoor Pursuit Centre of New Zealand: Mangatepopo Gorge incident, 15th April 2008.

Burke, S.C. (2004) Team task analysis. In: N.A. Stanton, A. Hedge, K. Brookhuis, E. Salas and H. Hendrick (eds), *Handbook of Human Factors and Ergonomics Methods*. Boca Raton, USA: CRC Press, pp. 56.1–56.8.

Cannon-Bowers, J., Tannenbaum, S., Salas, E. and Volpe, C. (1995) Defining competencies and establishing team training requirements. In: R. Guzzo, and E. Salas (eds), *Team Effectiveness and Decision-making in Organisations*. San Francisco: Jossey Bass.

Card, S.K., Moran, T.P. and Newell, A. (1983) *The Psychology of Human-Computer Interaction*. Hillsdale, NJ: Lawrence Erlbaum Associates.

Cassano-Piche, A.L., Vicente, K.J. and Jamieson, G.A. (2009) A test of Rasmussen's risk management framework in the food safety domain: BSE in the UK. *Theoretical Issues in Ergonomics Science*, 10:4, 283–304.

Cawley, J.C. (2003) Electrical accidents in the mining industry, 1990–1999. IEEE *Transactions on Industry Applications*, 39, 1570–6.

Celik, M. and Cebi, S. (2009) Analytical HFACS for investigating human errors in shipping accidents. *Accident Analysis and Prevention*, 41:1, 66–75.

Celik, M., Lavasani, S.M. and Wang, J. (2010) A risk-based modelling approach to enhance shipping accident investigation. *Safety Science*, 48:1, 18–27.

Cheng, A.S-K. and Ng, T.C.-K. (2010) Development of a Chinese motorcycle rider driving violation questionnaire. *Accident Analysis and Prevention*, 42:4, 1250–56.

Choudhry, R.M. and Fang, D. (2008) Why operatives engage in unsafe work behavior: investigating factors on construction sites. *Safety Science*, 46:4, 566–84.

Clarke, D.D., Ward, P., Bartle, C. and Truman, W. (2010) Killer crashes: fatal road traffic accidents in the UK. *Accident Analysis and Prevention*, 42:2, 764–70.

Cooper, P. (2003) Coalition deaths fewer than in 1991: we became stronger while Saddam became weaker. <www.cnn.com>, accessed 25 June 2003.

Cox, T. and Cox, S. (1996) Work-related stress and control room operations in nuclear power generation. In: N. Stanton (ed.), *Human Factors in Nuclear Safety*. London, UK: Taylor and Francis pp. 251–76.

Crandall, B., Klein, G. and Hoffman, R. (2006) *Working Minds: A Practitioner's Guide to Cognitive Task Analysis*. Cambridge, MA: MIT Press.

Cullen, Rt Hon Lord (2000) *The Ladbroke Grove Rail Inquiry. Part 1 Report*. Norwich: HSE Books, HMSO.

Dekker, S.W.A. (2002) Reconstructing human contributions to accidents: the new view on human error and performance. *Journal of Safety Research*, 33, 371–85.

Denis, D., St-Vincent, M., Imbeau, D. and Trudeau, R. (2006) Stock management influence on manual materials handling in two warehouse superstores. *International Journal of Industrial Ergonomics*, 36:3, 191–201.

Diaper, D. and Stanton, N.S. (2004) *The Handbook of Task Analysis for Human-computer Interaction*. New Jersey: Lawrence Erlbaum Associates.

Doytchev, D.E. and Szwillus, G. (2009) Combining task analysis and fault tree analysis for accident and incident analysis: a case study from Bulgaria. *Accident Analysis and Prevention*, 41:6, 1172–9.

Driskell, J.E. and Mullen, B. (2004) Social network analysis. In: N.A. Stanton, A. Hedge, K. Brookhuis, E. Salas and H. Hendrick (eds), *Handbook of Human Factors and Ergonomics Methods*. Boca Raton, USA: CRC Press, pp. 58.1–58.6.

El Bardissi, A.W., Wiegmann, D.A., Dearani, J.A., Daly, R.C. and Sundt, T.M. (2007) Application of the human factors analysis and classification system methodology to the cardiovascular surgery operating room. *Annals of Thoracic Surgery*, 83, 1412–18.

Embrey, D.E. (1986) SHERPA: A Systematic Human Error Reduction and Prediction Approach. Paper presented at the International Meeting on Advances in Nuclear Power Systems. Knoxville, Tennessee, USA.

Endsley, M.R. (1995) *Towards a Theory of Situation Awareness in Dynamic Systems*. Human Factors, 37, pp. 32–64.

REFERENCES

Eysenck, M.W. and Keane, M.T. (1990) *Cognitive Psychology: A Student's Handbook*. Hove: Lawrence Erlbaum.

Feyer, A-M. Williamson, A.M., Stout, N., Driscoll, T., Usher, H. and Langley, J. (2001) Comparison of work-related fatal injuries in the United States, Australia and New Zealand: method and overall findings. *Injury Prevention*, 7, 22–8.

Finch, C., Boufous, S. and Dennis, R. (2007) Sports and leisure injury hospitalisations in NSW, 2003–2004: sociodemographic and geographic patterns and sport-specific profiles. NSW Injury Risk Management Research Centre.

Flanagan, J.C. (1954) The critical incident technique. *Psychological Bulletin*, 51:4, 327–58.

Flores, A., Haileyesus, T. and Greenspan, A. (2008) National estimates of outdoor recreational injuries treated in emergency departments, United States, 2004–2005. *Wilderness and Environmental Medicine*, 19, 91–8.

Gaur, D. (2005) Human factors analysis and classification system applied to civil aircraft accidents in India. *Aviation Space and Environmental Medicine*, 76, pp. 501–5.

Getty, D.J., Swets, J.A., Pickett, R.M. and Gonthier, D. (1995) System operator response to warnings of danger: a laboratory investigation to the effects of the predictive value of a warning on human response time. *Journal of Experimental Psychology: Applied*, 1:1, 19–33.

Gorman, J.C., Cooke, N. and Winner, J.L. (2006) Measuring team situation awareness in decentralised command and control environments. *Ergonomics*, 49:12–13, 1312–26.

Grabowski, M., You, Z., Zhou, Z., Song, H., Steward, M. and Steward, B. (2009) Human and organizational error data challenges in complex, large scale systems. *Safety Science*, 47:8, 1185–94.

Graham, R. (1999) Use of auditory icons as emergency warnings: evaluation within a vehicle collision avoidance application. *Ergonomics*, 42:9, 1233–48.

Gray, W.D., John, B.E. and Atwood, M.E. (1993) Project ernestine: validating a GOMS analysis for predicting and explaining real-world performance. *Human-Computer Interaction*, 8, 237–309.

Griffin, T.G.C., Young, M.S. and Stanton, N.A. (2010) Investigating accident causation through information network modelling. *Ergonomics*, 53:2, 198–210.

Groves, W., Kecojevic, V. and Komljenovic, D. (2007) Analysis of fatalities and injuries involving mining equipment. *Journal of Safety Research*, 38, 461–70.

Harms-Ringdahl, L. (2009) Analysis of safety functions and barriers in accidents. *Safety Science*, 47:3, 353–63.

Harris, D. and Li, W. (2011) An extension of the human factors analysis and classification system for use in open systems. *Theoretical Issues in Ergonomics Science*, 12:2, 108–28.

Harris, D., Stanton, N.A., Marshall, A., Young, M.S., Demagalski, J. and Salmon, P.M. (2005) Using SHERPA to predict design induced error on the flight deck. *Aerospace Science and Technology*, 9:6, 525–32.

Health and Safety Executive (1999) *Reducing Error and Influencing Behaviour* HSE Books ISBN 0 717624528.

Health and Safety Executive (2008) Human factors: procedures. <http://www.hse.gov.uk/humanfactors/comah/procedures.htm>.

Heinrich, H.W. (1931) *Industrial Accident Prevention: A Scientific Approach*. New York: McGraw-Hill.

Hobbs, A. and Williamson, A. (2002) Unsafe acts and unsafe outcomes in aircraft maintenance. *Ergonomics*, 45:12, 866–82.

Hollnagel, E. (1998) *Cognitive Reliability and Error Analysis Method – CREAM*. 1st edition, Oxford: Elsevier Science.

Hollnagel, E. (2004) *Barriers and Accident Prevention*. Aldershot, UK: Ashgate Publishing.

Hone, K. and Baber, C. (2001) Designing habitable dialogues for speech-based interaction with computers. *International Journal of Human Computer Studies*, 54, 637–62.

Hopkins, A. (2000) *Lessons from Longford: The Esso Gas Plant Explosion*. Sydney: CCH.

Houghton, R.J., Baber, C., McMaster, R., Stanton, N.A., Salmon, P.M., Stewart, R. and Walker, G.H. (2006) Command and control in emergency services operations: a social network analysis. *Ergonomics*, 49:12–13, 1204–25.

Houghton, R.J., Baber, C., Cowton, M., Stanton, N.A. and Walker, G.H. (2008) WESTT (Workload, Error, Situational Awareness, Time and Teamwork): an analytical prototyping system for command and control. *Cognition Technology and Work*, 10:3, 199–207.

Iden, R. and Shappell, S.A. (2006) A human error analysis of U.S. fatal highway crashes 1990–2004. Paper presented at the Human Factors and Ergonomics Society 50th Annual Meeting, Santa Monica, CA.

Independent Police Complaints Commission (IPCC) (2007) Stockwell One: investigation into the shooting of Jean Charles de Menezes at Stockwell underground station on 22 July 2005. Published by the Independent Police Complaints Commission (IPCC) on 8 November 2007.

Jacinto, C., Canoa, M. and Guedes Soares, C. (2009) Workplace and organisational factors in accident analysis within the food industry. *Safety Science*, 47:5, 626–35.

Jenkins, D.P., Salmon, P.M., Stanton, N.A. and Walker, G.H. (2010) A systemic approach to accident analysis: a case study of the Stockwell shooting. *Ergonomics*, 3:1, 1–17.

Jenkins, S. and Jenkinson, P. (1993) Report into the Lyme Bay canoe tragedy. Devon County Council report.

John, B.A. and Newell, A. (1990) Toward an engineering model of stimulus-response compatibility. In: R.W. Proctor and T.G. Reeve (eds), *Stimulus-response Compatibility*. Amsterdam: North-Holland Publishing, pp. 427–79.

Johnson, C.W. and de Almeida, I.M. (2008) Extending the borders of accident investigation: applying novel analysis techniques to the loss of the Brazilian space launch vehicle VLS-1 V03. *Safety Science*, 46:1, 38–53.

Karra, V.K. (2005) Analysis of non-fatal and fatal injury rates for mine operator and contractor employees and the influence of work location. *Journal of Safety Research*, 36:5, 413–21.

Kecojevic, V., Komljenovic, D., Groves, W. and Radomsky, M. (2007) An analysis of equipment-related fatal accidents in U.S. mining operations: 1995–2005. *Safety Science*, 45, 864–74.

Kirwan, B. (1992a) Human error identification in human reliability assessment. Part 1: overview of approaches. *Applied Ergonomics*, 23:5, 299–318.

Kirwan, B. (1992b) Human error identification in human reliability assessment. Part 2: detailed comparison of techniques. *Applied Ergonomics*, 23:6, 371–81.

Kirwan, B. (1998a) Human error identification techniques for risk assessment of high-risk systems. Part 1: review and evaluation of techniques. *Applied Ergonomics*, 29:3, 157–77.

Kirwan, B. (1998b) Human error identification techniques for risk assessment of high-risk systems. Part 2: towards a framework approach. *Applied Ergonomics*, 29:5, 299–319.

Kirwan, B. and Ainsworth, L.K. (1992) *A Guide to Task Analysis*. London, UK: Taylor and Francis.

Klein, G. and Armstrong, A.A. (2004) Critical decision method. In: N.A. Stanton, A. Hedge, E. Salas, H. Hendrick and K. Brookhaus (eds), *Handbook of Human Factors and Ergonomics Methods*. Boca Raton, FL: CRC Press, pp. 35.1–35.8.

Klein, G.A., Calderwood, R. and Clinton-Cirocco, A. (1986) Rapid decision-making on the fireground. Proceedings of the 30th Annual Human Factors Society Conference. Dayton, OH: Human Factors Society, pp. 576–80.

Klein, G., Calderwood, R, and McGregor, D. (1989) Critical decision method for eliciting knowledge. *IEEE Transactions on Systems, Man and Cybernetics*, 19:3, 462–72.

Klinger, D.W. and Hahn, B.B. (2004) Team decision requirement exercise: making team decision requirements explicit. In: N.A. Stanton, A. Hedge, K. Brookhuis, E. Salas and H. Hendrick (eds), *Handbook of Human Factors Methods*. Boca Raton: CRC Press.

Kontogiannis, T., Kossiavelou, Z. and Marmaras, N. (2002) Self-reports of aberrant behaviour on the roads: errors and violations in a sample of Greek drivers. *Accident Analysis and Prevention*, 34, 381–99.

Kowalski-Trakofler, K. and Barrett, E. (2007) Reducing non-contact electric arc injuries: an investigation of behavioral and organizational issues. *Journal of Safety Research*, 38, 597–608.

Lawton, R. (1998) Not working to rule: understanding procedural violations at work. *Safety Science*, 28:2, 77–95.

Lawton, R. and Ward, N.J. (2005) A systems analysis of the Ladbroke Grove rail crash. *Accident Analysis and Prevention*, 37, 235–44.

Le Coze, J.-C. (2010) Accident in a French dynamite factory: an example of an organisational investigation. *Safety Science*, 48:1, 80–90.

Leigh, J.P., Waehrer, G., Miller, T.R. and Keenan, C. (2004) Costs of occupational injury and illness across industries. *Scandinavian Journal of Work, Environment, and Health*, 30, 199–205.

Lenné, M., Salmon, P., Regan, M., Haworth, N. and Fotheringham, N. (2007) Using aviation insurance data to enhance general aviation safety: phase 1 feasibility study. In: J.M. Anca (ed.), *Multimodal Safety Management and Human Factors – Crossing the Borders of Medical, Aviation, Road and Rail Industries*. Aldershot, UK: Ashgate Publishing, pp. 73–82.

Lenné, M.G., Ashby, K. and Fitzharris, M. (2008) Analysis of general aviation crashes in Australia using the human factors analysis and classification system. *International Journal of Aviation Psychology*, 18, 340–52.

Leveson, N.G. (2002) *A New Approach to System Safety Engineering*. Cambridge, MA: Aeronautics and Astronautics, Massachusetts Institute of Technology.

Leveson, N.G. (2004) A new accident model for engineering safer systems. *Safety Science*, 42:4, 237–70.

Leveson, N.G. (2009) The need for new paradigms in safety engineering. Safety-critical systems: problems, process and practice: 3-20. Proceedings of the Seventeenth Safety-critical Systems Symposium, Brighton, UK, 3–5 February 2009.

Leveson, N., Allen, P. and Storey, M.A. (2002) The analysis of a friendly fire accident using a systems model of accidents. In: *Proceedings of the 20th International System Safety Society Conference (ISSC 2003)*. Unionville, VA: System Safety Society, pp. 345–57.

Leveson, N., Daouk, M., Dulac, N. and Marias, K. (2003) A systems theoretic approach to safety engineering. Paper presented at workshop on investigation and reporting of incidents and accidents (IRIA), 16–19 September 2003, Virginia, USA.

Li, W.-C. and Harris, D. (2006) Pilot error and its relationship with higher organizational levels: HFACS analysis of 523 accidents. *Aviation, Space and Environmental Medicine*, 77, 1056–61.

Li, W.-C., Harris, D. and Yu, C.-S. (2008) Routes to failure: analysis of 41 civil aviation accidents from the Republic of China using the human factors analysis and classification system. *Accident Analysis and Prevention*, 40:2, 426–34.

Lombardi, D.A., Verma, S.K., Brennan, M.J. and Perry, M.J. (2009) Factors influencing worker use of personal protective eyewear. *Accident Analysis and Prevention*, 41:4, 755–62.

Maiti, J. and Bhattacherjee, A. (1999) Evaluation of risk of occupational injuries among underground coal mine workers through multinomial logit analysis. *Journal of Safety Research*, 30, 93–101.

Marsden, P. (1996) Procedures in the nuclear industry. In: N. Stanton (ed.), *Human Factors in Nuclear Safety*. London, UK: Taylor and Francis, pp. 99–116.

McKinsey and Company (2002a) Improving NYPD emergency preparedness and response. August 19, 2002.

McKinsey and Company (2002b) FDNY Fire operations response on September 11. McKinsey and Company report. <http://www.nyc.gov/html/fdny/html/mck_report/toc.html>.

Metropolitan Police Service (MPS) (2008) Suspected suicide bombers – Operation Kratos. <http://www.met.police.uk/docs/kratos_briefing.pdf>, accessed 12/02/2009.

Militello, L.G. and Hutton, J.B. (2000) Applied Cognitive Task Analysis (ACTA): A practitioner's toolkit for understanding cognitive task demands. In: J. Annett and N.S. Stanton (eds), *Task Analysis*. London, UK: Taylor and Francis, pp. 90–113.

Ministry of Defence (2004) Army board of inquiry report. <http://www.mod.uk/NR/rdonlyres/C2384518-7EBA-4CFF-B127-E87871E41B51/0/boi_challenger2_25mar03.pdf>, accessed December 2007.

Murrell, K.F.H. (1965) *Human Performance in Industry*. New York: Reinhold Publishing.

Nallet, N., Bernard, M. and Chiron, M. (2010) Self-reported road traffic violations in France and how they have changed since 1983. *Accident Analysis and Prevention*, 42:4, 1302–9.

NATO (2002) Code of best Practice for C2 assessment. Department of Defense Command and Control Research Program (CCRP), 3rd Edition.

Nelson, P.S. (2008) A STAMP analysis of the LEX COMAIR 5191 accident. Unpublished MSc thesis, <http://sunnyday.mit.edu/papers/nelson-thesis.pdf>, accessed 27 Jan 2011.

Nivolianitou, Z.S., Leopoulos, V.N. and Konstantinidou, M. (2004) Comparison of techniques for accident scenario analysis in hazardous systems. *Journal of Loss Prevention in the Process Industries*, 17:6, 467–75.

O'Hare, D., Wiggins, M., Williams, A. and Wong, W. (2000) Cognitive task analysis for decision centred design and training. In: J. Annett and N.A. Stanton (eds), *Task Analysis*. London: Taylor and Francis, pp. 170–90.

Ockerman, J. and Pritchett, A. (2001) A review and reappraisal of task guidance: aiding workers in procedure following. *International Journal of Cognitive Ergonomics*, 4:3, 191–212.

Olsen, J.R. and Nielsen, E. (1988) The growth of cognitive modeling in human-computer interaction since GOMS. *Human-Computer Interaction*, 3, 309–50.

Olsen, J.R. and Olsen, G.M. (1990) The growth of cognitive modeling in human-computer interaction since GOMS. *Human-Computer Interaction*, 5, 221–65.

Olsen, N.S. and Shorrock, S.T. (2010) Evaluation of the HFACS-ADF safety classification system: inter-coder consensus and intra-coder consistency. *Accident Analysis and Prevention*, 42:2, 437–44.

Paletz, S.B.F., Bearman, C., Orasanu, J. and Holbrook, J. (2009) Socializing the human factors analysis and classification system: incorporating social psychological phenomena into a human factors error classification system. *Human Factors*, 51:4, 435–45.

Patterson, E.S., Rogers, M.L., Chapman, R.J. and Render, M.L. (2006) Compliance with intended use of barcode medication administration in acute and long term care: an observational study. *Human Factors*, 48:1, 15–22.

Patterson, J.M. and Shappell, S.A. (2010) Operator error and system deficiencies: analysis of 508 mining incidents and accidents from Queensland, Australia using HFAC. *Accident Analysis and Prevention*, 42:4, 1379–85.

Paul, P.S. and Maiti, J. (2008) The synergic role of sociotechnical and personal characteristics on work injuries in mines. *Ergonomics*, 51, 737–67.

Perrow, C. (1999) *Normal Accidents: Living With High-risk Technologies*. Princeton, New Jersey: Princeton University Press.

Piper, J.L. (2001) *A Chain of Events: The Government Cover-up of the Black Hawk Incident and the Friendly-fire Death of Lt. Laura Piper*. Virginia: Brasseys, Inc.

Rafferty, L.A., Stanton, N.A. and Walker, G.H. (2010) The famous five factors in teamwork: a case study of fratricide. *Ergonomics*, 53:10, 1187–1204.

Rafferty, L.A., Stanton, N.A. and Walker, G.H. (2009) FEAST: Fratricide Event Analysis of Systemic Teamwork. *Theoretical Issues in Ergonomics Science*.

Rafferty, L.A., Stanton, N. A. and Walker, G. H. (In Press) *Human Factors of Fratricide*. Aldershot, UK: Ashgate Publishing

Rashid, H.S.J., Place, C.S. and Braithwaite, G.R. (2010) Helicopter maintenance error analysis: beyond the third order of the HFACS-ME. *International Journal of Industrial Ergonomics*, 40:6, 636–47.

Rasmussen, J. (1997) Risk management in a dynamic society: a modelling problem. *Safety Science*, 27:2/3, 183–213.

Reason, J. (1990) *Human Error*. Cambridge: Cambridge University Press.

Reason, J. (1995) A Systems Approach to Organizational Error. *Ergonomics*, 38, 1708–21.

Reason, J. (1997) *Managing the Risks of Organisational Accidents*. Burlington, VT: Ashgate Publishing.

Reason, J. (2002) Error management: combating omission errors through task analysis and good reminders. *Quality and Safety in Health Care*, 11:1, 40–44.

Reason, J. (2008) *The Human Contribution: Unsafe Acts, Accidents and Heroic Recoveries*. Aldershot, UK: Ashgate Publishing.

Reinach, S. and Viale, A. (2006) Application of a human error framework to conduct train accident/incident investigations. *Accident Analysis and Prevention*, 38, 396–406.

Riley, J. and Meadows, J. (1995) The role of information in disaster planning: a case study approach. *Library Management*, 16:4, 18–24.

Ripley, T. (2003) Combatting friendly fire. <www.ft.com>, accessed 23 January 2003.

Royal Australian Air Force (2001) The report of the F-111 deseal/reseal board of inquiry. Canberra, ACT: Air Force Head Quarters.

Salas, E., Sims, D.E. and Burke, C.S. (2005) Is there a big five in teamwork? *Small Group Research*, 36:5, 555–99.

Salmon, P.M., Stanton, N.A., Walker, G.H., Jenkins, D.P., Baber, C. and McMaster, R. (2008) Representing situation awareness in collaborative systems: a case study in the energy distribution domain. *Ergonomics*, 51:3, 367–84.

Salmon, P.M., Stanton, N.A., Walker, G.H. and Jenkins, D.P. (2009) *Distributed Situation Awareness: Advances in Theory, Measurement and Application to Teamwork*. Aldershot, UK: Ashgate Publishing.

Salmon, P.M., Lenné, M.G. and Stephan, K. (2010) Applying systems based methods to road traffic accident analysis: the barriers to application in an open, unregulated system. 9th International Symposium of the Australian Aviation Psychology Association, 18–22 April, Sydney.

Salmon, P.M., Stanton, N.A., Gibbon, A.C., Jenkins, D.P. and Walker, G.H. (2010) *Human Factors Methods and Sports Science: A Practical Guide*. Boca Raton, USA: Taylor and Francis.

Salmon, P.M., Williamson, A., Lenné, M.G., Mitsopoulos, E. and Rudin-Brown, C.M. (2010) Systems-based accident analysis in the led outdoor activity domain: application and evaluation of a risk management framework. *Ergonomics*, 53:8, 927–39.

Schraagen, J.M., Chipman, S.F. and Shalin, V.L. (2000) *Cognitive Task Analysis*. Mahwah, NJ: Lawrence Erlbaum Associates.

Shadbolt, N.R. and Burton, M. (1995) Knowledge elicitation: a systemic approach. In: J.R. Wilson and E.N. Corlett (eds), *Evaluation of Human Work: A Practical Ergonomics Methodology*. London: Taylor and Francis, pp. 406–40.

Shappell, S.A. and Wiegmann, D.A. (1997) A human error approach to accident investigation: the taxonomy of unsafe operations. *The International Journal of Aviation Psychology*, 7:4, 269–91.

Shappell, S.A. and Wiegmann, D.A. (2001) Unravelling the mystery of general aviation controlled flight into terrain accidents using HFACS. Paper presented at the 11th International Symposium on Aviation Psychology. Columbus, OH: The Ohio State University.

Shappell, S.A. and Wiegmann, D.A. (2003) A human error analysis of general aviation controlled flight into terrain accidents occurring between 1990–1998. Report number DOT/FAA/AM-03/4. Washington DC: Federal Aviation Administration.

Shappell, S.A., Detwiler, C., Holcomb, K., Hackworth, C., Boquet, A. and Wiegmann, D.A. (2007) Human error and commercial aviation accidents: an analysis using the human factors analysis and classification system. *Human Factors*, 49, 227–42.

Shi, J., Bai, Y., Ying, X. and Atchley, P. (2010) Aberrant driving behaviors: a study of drivers in Beijing. *Accident Analysis and Prevention*, 42:4, 1031–40.

Shorrock, S.T. (2006) Technique for the retrospective and predictive analysis of human error (TRACEr and TRACEr-lite). In: W. Karwowski (ed.), *International Encyclopedia of Ergonomics and Human Factors*. 2nd Edition, London: Taylor and Francis.

Shorrock, S.T. and Kirwan, B. (2002) Development and application of a human error identification tool for air traffic control. *Applied Ergonomics*, 33, 319–36.

Skillicorn, D.B. (2004) Social network analysis via matrix decompositions: Al-Qaeda. Unpublished manuscript.

Snook, S.A. (2000) *Friendly Fire: The Accidental Shootdown of US Black Hawks Over Northern Iraq*. Princeton, NJ: Princeton University Press.

Stanton, N.A. (2006) Hierarchical task analysis: developments, applications, and extensions. *Applied Ergonomics*, 37:1, 55–79.

Stanton N.A. and Baber, C. (2008) Modelling of human alarm handling responses times: a case of the Ladbroke Grove rail accident in the UK. *Ergonomics*, 51:4, 423–40.

Stanton, N.A. and Young, M. (1999) *A Guide to Methodology in Ergonomics: Designing for Human Use*. London: Taylor and Francis.

Stanton, N.A., Hedge, A., Brookhuis, K., Salas, E. and Hendrick, H. (2004) *Handbook of Human Factors Methods*. Boca Raton, USA: CRC Press.

Stanton, N.A., Salmon, P.M., Walker, G., Baber, C. and Jenkins, D.P. (2005) *Human Factors Methods: A Practical Guide for Engineering and Design*. Aldershot, UK: Ashgate Publishing.

Stanton, N.A., Stewart, R., Harris, D., Houghton, R.J., Baber, C., McMaster, R., Salmon, P.M., Hoyle, G., Walker, G.H., Young, M.S., Linsell, M., Dymott, R. and Green, D. (2006) Distributed situation awareness in dynamic systems: theoretical development and application of an ergonomics methodology. *Ergonomics*, 49, 1288–1311.

Stanton, N.A., Walker, G.H., Young, M.S., Kazi, T.A. and Salmon, P. (2007) Changing drivers' minds: the evaluation of an advanced driver coaching system. *Ergonomics*, 50:8, 1209–34.

Stanton, N.A., Baber, C. and Harris, D. (2008) *Modelling Command and Control: Event Analysis of Systemic Teamwork*. Aldershot, UK: Ashgate.

Stanton, N.A., Salmon, P.M., Harris, D., Demagalski, J., Marshall, A., Young, M.S., Dekker, S.W.A. and Waldmann, T. (2009) Predicting pilot error: testing a new method and a multi-methods and analysts approach. *Applied Ergonomics*, 40:3, 464–71.

Stanton, N.A., Jenkins, D.P., Salmon, P.M., Walker, G.H., Rafferty, L. and Revell, K. (2010) *Digitising Command and Control: A Human Factors and Ergonomics Analysis of Mission Planning and Battlespace Management*. Aldershot, UK: Ashgate Publishing.

Stanton, N.A., Salmon, P.M., Walker, G.H. and Jenkins, D.P. (2010) *Human Factors and the Design and Evaluation of Central Control Room Operations*. Boca Raton, USA: Taylor and Francis.

Stewart, R., Stanton, N.A., Harris, D., Baber, C., Salmon, P.M., Mock, M., Tatlock, K., Wells, L. and Kay, A. (2008) Distributed situation awareness in an airborne warning and control system: application of novel ergonomics methodology. *Cognition Technology and Work*, 10:3, 221–9.

St-Vincent, M., Denis, D., Imbeau, D. and Laberge, M. (2005) Work factors affecting manual materials handling in a warehouse superstore. *International Journal of Industrial Ergonomics*, 35:1, 33–46.

Svedung, I. and Rasmussen, J. (2002) Graphic representation of accident scenarios: mapping system structure and the causation of accidents. *Safety Science*, 40:5, 397–417.

Taylor, Lord Justice (1990) The Hillsborough Stadium disaster: final report. London: HMSO.

United States General Accounting Office (1997) Operation Provide Comfort: review of U.S. Air Force investigation of Black Hawk fratricide incident. Report to Congressional requesters.

USAF Aircraft Accident Investigation Board (2004) U.S. Army Black Hawk helicopters 87-26000 and 88- 645 26060: Vol. 1, executive summary: UH-60 Black Hawk helicopter accident. 14 April 1994 online, <www.schwabhall.com/opc report.htm>.

Van Duijn, M.A.J. and Vermunt, J.K. (2006) What is special about social network analysis? *Methodology*, 2:1, 2–6.

Vicente, K.J. (1999) *Cognitive Work Analysis: Toward Safe, Productive, and Healthy Computer-based Work*. Mahwah, NJ: Lawrence Erlbaum Associates.

Vicente, K.J. and Christoffersen, K. (2006) The Walkerton E. Coli outbreak: a test of Rasmussen's framework for risk management in a dynamic society. *Theoretical Issues in Ergonomics Science*, 7:2, 93–112.

Wagenaar, W.A. and Reason, J.T. (1990) Types and tokens in road accident causation. *Ergonomics*, 33, 1365–75.

Walker, D. (2007) Applying the Human Factors Analysis and Classification System (HFACS) to incidents in the UK construction industry. Unpublished MSc thesis, Cranfield University, Bedford, UK.

Walker, G.H., Gibson, H., Stanton, N.A., Baber, C., Salmon, P.M. and Green, D. (2006) Event analysis of systemic teamwork (EAST): a novel integration of ergonomics methods to analyse C4i activity. *Ergonomics*, 49, 1345–69.

Walker, G.H., Stanton, N.A., Kazi, T.A., Salmon, P.M. and Jenkins, D.P. (2009) Does advanced driver training improve situation awareness? *Applied Ergonomics*, 40:4, 678–87.

Walker, G.H., Stanton, N.A., Baber, C., Wells, L., Jenkins, D.P. and Salmon, P.M. (2010) From ethnography to the EAST method: a tractable approach for representing distributed cognition in air traffic control. *Ergonomics*, 53:2, 184–97.

Walker, G.H., Stanton, N.A. and Salmon, P.M. (2011) Cognitive compatibility of motorcyclists and car drivers. *Accident Analysis and Prevention*, 43:3, 878–88.

Ward, R., Brazier, A. and Lancaster, R. (2004) Different types of supervision and the impact on safety in the chemical and allied industries: literature review. Health and safety executive report.

Watts, L.A. and Monk, A.F. (2000) Reasoning about tasks, activities and technology to support collaboration. In: J. Annett and N. Stanton (eds), *Task Analysis*. London, UK: Taylor and Francis, pp. 55–78.

Wenner, C.A. and Drury, C.G. (2000) Analysing human error in aircraft ground damage incidents. *International Journal of Industrial Ergonomics*, 26, 177–99.

Wiegmann, D.A. and Shappell, S.A. (2001) A human error analysis of commercial aviation accidents using the human factors analysis and classification system (HFACS). Report number DOT/FAA/AM-01/03. Washington DC: Federal Aviation Administration.

Wiegmann, D.A. and Shappell, S.A. (2003) *A Human Error Approach to Aviation Accident Analysis. The Human Factors Analysis and Classification System*. Burlington, VT: Ashgate Publishing Ltd.

Wikipedia (2010) Why-because analysis. <http://en.wikipedia.org/wiki/Why-Because_analysis>, accessed 11 February 2011.

Wilson, K.A., Salas, E., Priest, H.A. and Andrews, D. (2007) Errors in the heat of battle: taking a closer look at shared cognition breakdowns through teamwork. *Human Factors*, 49:2, 243–56.

Woo, D.M. and Vicente, K.J. (2003) Sociotechnical systems, risk management, and public health: comparing the North Battleford and Walterton outbreaks. *Reliability Engineering and System Safety*, 80, 253–69.

Woods, D.D., Dekker, S., Cook, R., Johannesen, L. and Sarter, N., (1994) *Behind Human Error: Cognitive Systems, Computers and Hindsight*. Ohio: CSERIC.

Wu, W., Gibb, A.G.F. and Li, Q. (2010) Accident precursors and near misses on construction sites: an investigative tool to derive information from accident databases. *Safety Science*, 48:7, 845–58.

9/11 Commission (2004) The 9/11 Commission Report. <www.9-11commission.gov/report/911Report.pdf>.

Index

accident analysis xix, xx, xxi, 2, 7–9, 23, 34, 44–45, 51, 68, 77–78, 80, 83, 104, 151, 177, 179
accident causation xix, 1–2, 10, 102, 111, 123, 143, 177
　domino theory 2
　epidemiological models 2
　models xx, 2, 11
　sequential models 2
　systemic models 2
accident databases 14, 34, 70, 97, 101, 111
accident investigation xx, 16–17, 23, 34
accident prevention xix, 44
Accimap xxi, 2, 10–13, 23–28, 37–39, 40, 83–96, 152, 177–178
agent association matrix 48
allocation of functions analysis 9
analogy 19, 125
analyst
　analysts 9–10, 12–15, 20–21, 24–26, 33–38, 40–45, 48–49, 52, 58–61, 67, 71–73, 78–81, 102–106, 109, 124, 136, 160, 177–179
　classification 106
　hindsight 13, 25, 37
　judgement 25, 38, 44, 61, 71, 79
　training 13–14, 20, 26, 30, 35, 38, 45, 51, 60, 68, 75, 80, 102, 177, 179
assessment 18, 125
aviation
　air traffic control 15–16, 41, 70, 78
　Australian General Aviation 97–109
　aviation xix, xxi, 2, 12–17, 23, 33–34, 37, 41, 53, 57, 97–109, 116, 117, 123, 178

basis of choice 19, 125
blame culture 20

CDA *see* Coordination Demands Analysis
CDM *see* Critical Decision Method
centrality 59, 164–165
Challenger II incident 133–142
chemical processing 29
Chernobyl disaster 123
cognitive probes 17–21
cognitive reliability and error analysis method 75
cognitive task analysis xxi, 7, 12–14, 124

communication 94–95, 11–16, 152, 154, 157–158, 162, 165, 167, 178
　analysis 14
　failure 135
communications usage diagram 78–81, 153, 165, 174
complex sociotechnical systems ixx, 2, 4, 5, 9, 13, 16, 23, 48, 51, 57, 89, 96, 151, 152, 176
concept maps 60
conceptual method 19, 125
construction xix, 34, 131
control
　failures 41–44
　structure diagram 42, 46,
　theory 4
coordination demands analysis 78–80, 153, 165, 174–175
CPA *see* Critical Path Analysis
CREAM *see* Cognitive Reliability and Error Analysis Method
crew resource management 114–116, 120, 121
critical decision method xxi, 10–13, 16–21, 57, 60, 78–81, 123–132, 177–178
critical incident technique 20
critical path analysis xxii, 13–16, 65–70, 143–150, 177–179
CTA *see* Cognitive Task Analysis
CUD *see* Communications Usage Diagram
cue identification 18–21, 125–126

decision making xx, 12, 17–19, 78–80, 94, 123, 125, 129–130, 153, 167
decision requirements exercise 20
distraction 19, 35
documentation review 7, 11–15, 29, 41, 45, 51, 52, 78
dynamite production xix

EAST *see* Event Analysis of Systemic Teamwork
emergency services 17, 48, 52, 54, 78
energy distribution 78
error
　communication 94–95
　decision 35, 104, 107, 109, 114–115, 120–121
　prediction 70

external error modes 70
false alarms 146
human 2, 8, 19, 29, 68, 103, 111, 129
internal error modes 70,71
perceptual 35, 103
psychological error mechanisms 70,71
skill-based 35, 36, 104, 107, 109, 113–117, 120–121
task 70,71, 80
evaluation 7
event analysis of systemic teamwork 14,16, 48, 52, 77–82, 151–152, 158, 160, 175–176, 178
expectancy 18–21, 125–126
experience 19, 125

farming xix
fatigue 135
    mental 35
    physical 35, 117
fault tree analysis 12–14, 28–33, 178
food industry xix
fratricide 133, 160, 163, 165, 175–176
FTA *see* Fault Tree Analysis

gas plant explosions 23
generalisation 19, 125
goal specification 18–21, 125–126
guidance 19, 125

hardware failure 29
HCI *see* Human Computer Interaction
healthcare xix, 34, 116, 131
Herald of Free Enterprise ferry disaster 30, 61
hierarchical task analysis 13–15, 48, 57, 66–68, 75–80, 147, 153, 159–160, 165, 174–175
HFACS *see* Human Factor Analysis and Classification System
Hillsborough football stadium disaster 26
hindsight 25, 37
HTA *see* Hierarchical Task Analysis
human computer interaction 13, 65–67, 146
human factor analysis and classification system xxi, 2, 11–14, 33–41, 178
human factors xix, xx, xxi, xxii, 1, 5, 7, 9, 65, 78, 80, 84, 90, 107, 111, 1511–52
hydro power plant 29

ICAM *see* Incident Cause Assessment Method
incident cause assessment method 111, 121
influencing factors 18–22, 125–126
information 70, 125
    elements 136–141, 158, 168, 171–175
    integration 19
interface evaluation 8
intervention 19, 25
interviews 7, 11–17, 19–23, 29, 41, 45, 48, 51, 124

transcripts 18–19, 124, 129
video recording 18–21, 35

Kegworth air disaster 57
keystroke level model method 67–68

Ladbroke Grove incident xxii, 66, 143
led outdoor activity xxi, 45–46, 83–87
    canoeing 38, 83–86
    whitewater rafting 17
logic gates 29
Lyme Bay incident xxi, 38–40, 83–88

Mangatepopo Gorge incident 45–47
manual handling 124–132
maritime 34
medical illness 35, 113
mental models 19, 125
military xxii, 17, 34, 41, 48, 57, 78, 133–142, 151–175,
    command and control 52, 57, 94, 135
mining xix, 34, 109–121, 131

network
    density 50, 59, 162–165
    diameter 59
    knowledge networks 54, 78, 152
    semantic networks 60
    social network 77, 152, 162
    task network 77, 152
nuclear power 29–30

observation 7, 11–15, 45, 48, 57, 60, 78, 80, 124
odds ratio 37, 111, 114
Operation Provide Comfort incident 78, 154–175
operation sequence diagrams 78,80–81, 153, 174
options 18,22, 125–126
OR *see* odds ratio
organisational
    climate 116, 120, 121
    factors 71, 103–104, 114, 123
    processes 116–119, 129
OSDs *see* Operation Sequence Diagrams
outdoor education xix, 23, 84

performance
    shaping factors 70
    time modelling 9, 13
planning 135–137, 141
police xxi, 23
procedural failures 119–120
product design 9
propositional networks xxii, 52, 53, 57, 77, 78, 80–81, 133–142, 153, 160, 168, 175, 178
    analysis xxii, 15,16, 52, 57–65, 133–142, 168–173
public health 23, 41

INDEX

questionnaires 14, 45, 48, 51

rail xix, xxii, 16–17, 23, 34, 48, 57, 70, 78, 131, 143–150
Rassmussen's risk management framework 2–3, 11, 12, 84, 177–178
Reason, James xix
reliability 20, 26, 30, 37–38, 44–45, 49–51, 60–61, 68, 75, 81, 84, 102, 104, 111, 160, 177
    inter-rater reliability 36–37, 73, 106, 109, 111
retail xxi, xxii, 17–21, 123–132
retrospective incident analysis component 70
road transport xix, 17, 34, 57, 131
    driving 123
    motorcycling 123

safety 4
    constraints 41, 42
    critical systems 4, 70, 118, 123, 129
    culture xx, 124, 131
    management 4
scenario-based design 9
September 11, 2011 terrorism attack 52–56
shipping xix, 29
situation awareness xx, 8, 14–19, 35, 53, 56, 60, 79– 80, 125, 133–142, 153, 167–168, 172–173, 175, 178
    distributed 57–65
    methods 8
    modelling 8
SME *see* Subject Matter Experts
SNA *see* Social Network Analysis
social desirability bias 20
social network analysis 12–15, 48–56, 78–81, 153, 161–165, 174, 178
sociometric status 59, 162–165, 168
software
    Agna 15, 50–52, 61, 79, 81
    Leximancer 60–61
    Microsoft Visio 14–15, 24, 26, 30, 33, 45, 61, 68, 81
    NetDraw 51
    NodeXL 51–52
    Pajek 51–52
    R 112
    UCINET 51, 52
    WESTT 14–15, 50, 60–61
space vehicles 23

STAMP *see* Systems Theoretic Accident Modelling and Processes
Stockwell incident xxi, 83, 89–96
storyboarding 9
subject matter experts 10, 17–23, 25, 29, 35, 36, 41, 44, 48, 50, 57–59, 61, 71, 74, 78–80
supervision 2, 12, 33, 35, 71, 103, 107, 109, 113–118, 120–121
Swiss cheese model 2–5, 11–12, 12, 33, 37, 38, 118, 143
system design xx, 7, 9
Systematic Human Error Prediction and Reduction Approach 75
Systems Theoretic Accident Modelling and Processes 2–5, 10–15, 41–47, 177–178
systems theory 117, 120

task analysis methods 7, 14, 124
teams xx, 90–91, 114,
    cognition 78
    performance measures 8
teamwork xx, 8, 79–80, 167
    assessment 14
Technique for the Retrospective and Predictive Analysis of Cognitive Errors 15–16, 70–77, 178
Tenerife air disaster (1977) 49
terrorism 48, 52
timeline analysis methods 20
TRACEr *see* Technique for the Retrospective and Predictive Analysis of Cognitive Errors
training xx, 2, 19, 21, 25, 34–36, 39, 43, 46, 71, 85, 87, 90, 97–99, 104–105, 116–121, 125–131, 143, 157–158, 166
    manuals 11, 57

US Federal Aviation Administration 17

validity 20, 26, 30, 38, 45, 51, 61, 68, 75, 81, 84, 111, 151, 177
verbal protocol analysis 15
violations 120–123, 129, 131, 151
    deliberate 117–118, 123
    unintentional 117

walkthroughs 7, 11
water contamination incident 43
workload xx, 8, 19, 174
    mental 8